Pro DAX and Data Modeling in Power BI

Creating the Perfect Semantic Layer to Drive Your Dashboard Analytics

Adam Aspin

Apress®

Pro DAX and Data Modeling in Power BI: Creating the Perfect Semantic Layer to Drive Your Dashboard Analytics

Adam Aspin
STAFFORD, UK

ISBN-13 (pbk): 978-1-4842-8994-5
https://doi.org/10.1007/978-1-4842-8995-2

ISBN-13 (electronic): 978-1-4842-8995-2

Managing Director, Apress Media LLC: Welmoed Spahr
Acquisitions Editor: Jonathan Gennick
Development Editor: Laura Berendson
Coordinating Editor: Jill Balzano

Cover Photo by Marek Piwnicki on Unsplash

Distributed to the book trade worldwide by Springer Science+Business Media LLC, 1 New York Plaza, Suite 4600, New York, NY 10004. Phone 1-800-SPRINGER, fax (201) 348-4505, e-mail orders-ny@springer-sbm.com, or visit www.springeronline.com. Apress Media, LLC is a California LLC and the sole member (owner) is Springer Science + Business Media Finance Inc (SSBM Finance Inc). SSBM Finance Inc is a **Delaware** corporation.

For information on translations, please e-mail booktranslations@springernature.com; for reprint, paperback, or audio rights, please e-mail bookpermissions@springernature.com.

Apress titles may be purchased in bulk for academic, corporate, or promotional use. eBook versions and licenses are also available for most titles. For more information, reference our Print and eBook Bulk Sales web page at http://www.apress.com/bulk-sales.

Any source code or other supplementary material referenced by the author in this book is available to readers on GitHub (https://github.com/Apress). For more detailed information, please visit http://www.apress.com/source-code.

Printed on acid-free paper

Table of Contents

About the Author

 **Adam Aspin** is an independent business intelligence consultant based in the United Kingdom. He has worked with SQL Server for over 25 years. During this time, he has developed several dozen reporting and analytical systems based on the Microsoft data and analytics product suite.

A graduate of Oxford University, Adam began his career in publishing before moving into IT. Databases soon became a passion, and his experience in this arena ranges from dBase to Oracle, and Access to MySQL, with occasional sorties into the world of DB2. He is, however, most at home in the Microsoft universe when using the Microsoft data and analytics stack–both in Azure and on-premises.

Business intelligence has been Adam's principal focus for the last 20 years. He has applied his skills for a range of clients in the areas of finance, banking, utilities, leisure, luxury goods, and pharmaceuticals.

Adam has been a frequent contributor to SQLServerCentral.com and *Simple Talk* for many years. He is a regular speaker at events such as SQL Saturdays and SQLBits. A fluent French speaker, Adam has worked in France and Switzerland for many years.

Adam is the author of popular Apress books: *SQL Server 2012 Data Integration Recipes*; *Business Intelligence with SQL Server Reporting Services*; *High Impact Data Visualization in Excel with Power View, 3D Maps, Get & Transform and Power BI*; *Data Mashup with Microsoft Excel Using Power Query and M*; *Pro Dashboard Creation with Power BI*; and *Pro Power BI Theme Creation*.

About the Technical Reviewer

Ed Freeman has been a data engineer ever since graduating with a mathematics degree from UCL in 2017. Throughout his career, he has been implementing intelligent cloud and data platforms built on Azure for clients of all sizes and industries. Ed is particularly passionate about Power BI and Microsoft data platform services as a whole, and heads *Power BI Weekly*, a free weekly newsletter to help the community keep up with the latest and greatest developments from the Power BI ecosystem.

Acknowledgments

Writing a technical book can prove to be a daunting challenge. So I am all the more grateful for all the help and encouragement that I have received from so many friends and colleagues.

First, my heartfelt thanks go, once again, to Jonathan Gennick, the commissioning editor of this book. Throughout the publication process, Jonathan has been an exemplary mentor. He has shared his knowledge and experience selflessly and courteously and provided much valuable guidance.

My deepest gratitude goes yet again to the Apress coordinating team for managing this volume through the rocks and shoals of the publication process.

When delving into the arcane depths of technical products, it is all too easy to lose sight of the main objectives of a book. Fortunately my good friend Ed Freeman, the technical reviewer, has worked unstintingly to help me retain focus on the objectives of this book. He has also shared his considerable experience of Power BI and DAX in the enterprise and has helped me immensely with his comments and suggestions.

Finally my deepest gratitude has to be reserved for the two people who have given the most to this book. They are my wife and son, who have always encouraged me to persevere while providing all the support and encouragement that anyone could want. I am very lucky to have both of them.

Introduction

Business intelligence (BI) is a concept that has been around for many years. Until recently, it has too often been a domain reserved for large corporations with teams of dedicated IT specialists. All too frequently, this has meant developing complex solutions using expensive products on timescales that did not meet business needs. All this has changed with the advent of self-service business intelligence.

Now a user with a reasonable knowledge of Microsoft Office can leverage their skills to produce their own analyses with minimal support from central IT. Then they can deliver their insights to colleagues safely and securely via the cloud. This democratization has been made possible by a superb free product from Microsoft - Power BI Desktop – that revolutionizes the way in which data is discovered, captured, structured, and shaped so that it can be sliced, diced, chopped, queried, and presented in an interactive and intensely visual way.

Power BI Desktop provides you with the capability to analyze and present your data and to shape and deliver your results easily and impressively. All this can be achieved in a fraction of the time that it would take to specify, develop, and test a corporate solution. To cap it all off, self-service BI with Power BI Desktop lets you produce reports at a fraction of the cost of more traditional solutions, with far less rigidity and overhead.

The aim of this short book is to introduce the reader to a fundamental aspect of this brave new world of user-driven dashboard development - building the data model that you need to deliver dashboards. This book will not involve a complete tour of Power BI Desktop, however. The product is simply too vast for that. Consequently, this book concentrates on the less visible - but nonetheless essential - layer of Power BI that is the data model.

The data model is important because the disparate data sources that you use in Power BI need molding into a unified structure in Power BI desktop if you are to take advantage of all that this amazing piece of software has to offer. Once a basic model has been designed and implemented, it can then be extended using DAX - Power BI's built-in analytics language – to create powerful custom analytics to drive your data visualization. Combining the data model and custom calculations creates a powerful basis for dashboard delivery.

Much of the analytical power that Power BI delivers comes from DAX. This acronym (short for **D**ata **A**nalysis e**X**pressions) is, in reality, an extremely robust and versatile set of functions that you can apply to deliver custom analytics. This power, however, comes at the price of a certain complexity. So the aim of Chapters 4 through 18 in this book is to give you a structured and progressive introduction to the power and capabilities of this language.

You need to be aware that DAX can be daunting at first sight. So this book does not aim to solve every potential challenge or deliver technical pirouettes to impress. Instead it aims to give you a clear grounding in the language so that you are then free to build on the basics and develop your own DAX expressions to solve your own analytical challenges.

Although a basic knowledge of the MS Office suite will help, this book presumes that you have little or no knowledge of DAX or data modeling in Power BI Desktop. These are therefore explained from the ground up with the aim of providing the most complete coverage possible of the way in which they work together to deliver data models that drive dashboards. Hopefully if you read the book and follow the examples given, you will arrive at a level of practical knowledge and confidence that you can subsequently apply to your own data modeling requirements. This book should prove invaluable to business intelligence developers, MS Office power users, IT managers, and finance experts — indeed anyone who wants to deliver efficient and practical business intelligence to their colleagues. Whether your aim is to develop a proof of concept or to deliver a fully fledged BI system, this book can, hopefully, be your guide and mentor.

You can, if you wish, read this book from start to finish, as it is designed to be a progressive self-tutorial. However, as dashboard development with Power BI Desktop is the interaction of a set of interdependent BI techniques, the book is broken down into several separate areas that correspond to the various facets of the product. They are as follows:

- Chapters 1 through 3 introduce you to the Power BI data model and the concept of the semantic layer.

- Chapters 4 through 7 begin your journey into DAX with the concept (and practice) of adding calculated columns to data tables.

- Chapters 8 through 18 explain the core elements of DAX measures - the formulas that you add to deliver analytics from across the data model.

This book comes with a small sample dataset that you can use to follow the examples that are provided. It may seem paradoxical to use a tiny data sample when explaining a product suite that is capable of analyzing medium and large datasets. However, I prefer to use an extremely simplistic data structure so that the reader is free to focus on the essence of what is being explained, and not the data itself.

If you need to learn further aspects of the Power BI Desktop ecosystem, then two companion volumes are available:

- *Pro Data Mashup for Power BI*

- *Pro Power BI Dashboard Creation*

The first guides you through the process of ingesting, cleansing, and shaping source datasets to underpin your dashboard creation. The second shows you how to use the data that you have prepared and modeled to create stunning dashboards.

Inevitably, not every question can be answered and not every issue can be resolved in one book. I truly hope that I have answered many of the essential DAX and data modeling questions that you will face when developing dashboards. Equally I hope that I have provided ways of solving a reasonable number of the challenges that you may encounter. I wish you good luck in using Power BI Desktop to prepare and deliver your insights. And I sincerely hope that you have as much fun with it as I had writing this book.

—Adam Aspin

CHAPTER 1

Using Power BI Desktop to Create a Data Model

Power BI's ability to access and load data from a wide variety of sources is undeniably one of the keys to its success as a world leading analytics tool. Yet simply fetching a lot of data does not by itself make creating accurate and informative interactive dashboards possible even in Power BI.

To deliver actionable analytics from multiple tables, you have to mold the tables into a structured data model. At its simplest, a data model can be considered to be a group of tables that can be queried together as a coherent set. Essentially this means linking the source data tables together in a structured way. This structure can then underpin the analyses that you display in your dashboards.

To start you on your path to creating data models in Power BI Desktop, this chapter will take you through the core techniques that you need to apply to create a simple data model. You will discover how to take a collection of source data tables and join them to create a structured data model that will enable you to deliver information, insight, and analysis from the data in the tables. Later chapters will take you through some of the more conceptual considerations that you may require to push data modeling to a higher level.

To simplify the learning process, this chapter will concentrate on explaining the tools and techniques that are available in Power BI Desktop to fuse disparate data tables into a unified data model.

This chapter will cover

- The Power BI Desktop Model window that you use to create and edit a data model

- Establishing relationships between the tables so that Power BI Desktop understands how the data in one table is linked to the data contained in another table

1

A. Aspin, *Pro DAX and Data Modeling in Power BI*, https://doi.org/10.1007/978-1-4842-8995-2_1

Creating data models is both an art and a science. Creating complex data models that allow enterprise-level analytics can take a lot of practice. The actual models that you will need will depend on the reporting challenges that you need to solve. The first three chapters in this book aim to introduce you to the techniques that Power BI Desktop makes available to deliver a data model.

I am presuming that you are up to speed on data ingestion in Power BI Desktop and so will not explain any of the details of this process here. Should you need further details, then these are available in the companion volume *Pro Data Mashup with Power BI*.

All the sample files used in this chapter are available on the Apress website. Once downloaded they should be in the folder C:\PowerBiDesktopSamples. The download and installation process is described in Appendix A.

Data Modeling in the Power BI Desktop Environment

Before leaping into the detail of what can be done to create a data model from a set of independent source tables, I think that it is best if you first familiarize yourself with the modeling environment itself.

The Model Icon

Core data modeling is done in the Model window. You switch to this window by clicking on the Model icon on the top left of the Power BI Desktop window as shown in Figure 1-1.

Figure 1-1. *The Model icon*

Switching to Model view displays the Model window which itself contains all the tables that are currently available. This view is a high-level overview of all the data tables you will be using to create and update the data model. Model view is particularly powerful as it allows you to step back from the detail and look at the dataset as a whole. This high-level overview is exactly what you need, because now it is time to think in terms of overall structures rather than the nitty gritty.

The whole point of Model view is to let you get a good look at the *entire* dataset and manage the relationships between data tables. An added feature is that you can also modify the layout in order to see the relationships between tables more clearly.

The Model Window

If you load the sample file PrestigeCarsRawSource.pbix, the Model window looks like the one displayed in Figure 1-2.

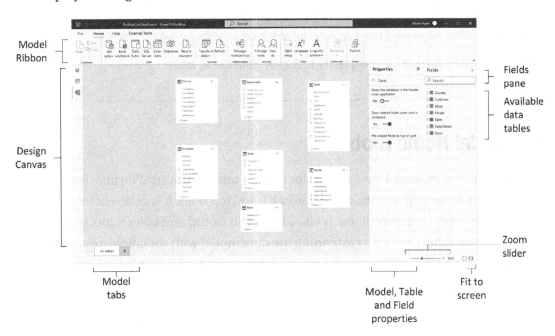

Figure 1-2. *The Model window*

The Model window contains the following elements.

Table 1-1. *The Model Window*

Element	Description
Home ribbon	The core ribbon containing modeling options
Design canvas	Contains all the available tables that can be used in the data model
Fields pane	The tables, fields, folders, and hierarchies that users will use to build dashboards
Properties pane	Additional data model properties that can be defined for • The data model itself • Tables • Fields
Data model tabs	Manage multiple views of the data model
Zoom slider	Allows you to zoom in or out of the actual data model
Fit to screen button	Adjusts the zoom factor to display all available tables

The Model Home Ribbon

The Home ribbon in Model view is used for loading and refreshing data just like the Home ribbon in the Report window of Power BI Desktop does. It is different to the Home ribbon of the Report window in that it also allows you to add and edit relationships (joins) between tables as well as extending the data model with specific attributes. The buttons that it contains are shown in Figure 1-3 and explained in Table 1-2.

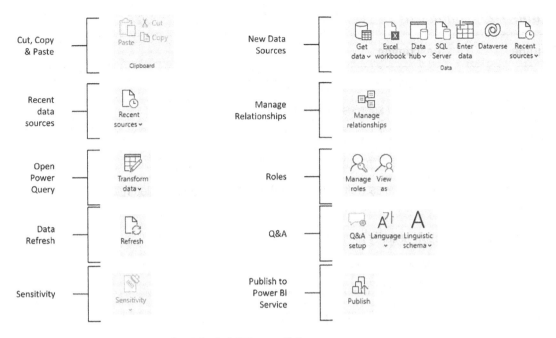

Figure 1-3. *Buttons in the Model Home ribbon*

Table 1-2. *The Modeling Ribbon Buttons*

Button	Description
Paste	Pastes data from the clipboard. Note that this cannot apply to tables, relationships, or fields in a data model
Cut	Cuts data to the clipboard. Note that this cannot apply to tables, relationships, or fields in a data model
Copy	Copies data to the clipboard. Note that this cannot apply to tables, relationships, or fields in a data model
Get Data	Add data from external sources to the Power BI data model
Excel Workbook	Add data from an Excel Workbook
Data hub	Add data from Power BI datasets and datamarts
SQL Server	Add data from a SQL Server database
Enter Data	Enter data manually into Power BI

(continued)

5

Table 1-2. (*continued*)

Button	Description
Dataverse	Add data from the Microsoft Dataverse
Recent sources	List recent sources of external data
Transform Data	Transform data using Power Query
Refresh	Refresh external data loaded into Power BI Desktop
Manage Relationships	Link tables as well as delete and modify these joins (called relationships)
Manage Roles	Apply security roles
View As	Test security roles
Q&A Setup	Setup AI Q&A in Power BI Desktop
Language	Set the language used for Q&A in Power BI Desktop
Linguistic Schema	Apply a Linguistic Schema used for Q&A in Power BI Desktop
Sensitivity	Set the Sensitivity (confidentiality) labels
Publish	Publish a dashboard to the Power BI Service

The only button on the Home button that we will be using in this chapter is Manage Relationships.

Designing a Power BI Desktop Data Model

Now that you have understood the Model window, you have started on the process to developing a high-performance data model for self-service business intelligence (BI). The next step, once you have imported data from one or several sources into the Power BI Desktop Data Model, is to ensure that your dataset is ready for initial use as a self-service BI data repository. This means creating and managing relationships between tables. This is a fundamental aspect of designing a structured and useable data model in Power BI Desktop.

When you initially load a set of tables into Power BI Desktop – and this is really important – the tables are generally *independent* of one another, and they cannot yet be queried as a coherent whole. In other words, while you can filter data in a single table, any filters or slicers based on one table will have *no effect on data in any other table*. Consequently, what matters now is to enable the tables to work together almost as if you were using one vast single table (but much more efficiently than is possible when using a single table). Enabling separate tables to work together is done through establishing relationships between tables.

Note   As establishing relationships between tables is so fundamental, Power BI defaults to attempting to join tables if it can. Sometimes this approach can do most of the work for you. In other cases it causes more problems than it is worth and you have to undo the joins that it created. Inhibiting and enabling this functionality are explained a little later in this chapter.

Before leaping into the technicalities of managing table relationships, we first need to answer a couple of simple questions:

- What are relationships between tables?

- Why do we need them?

Table relationships are links (or joins if you prefer) between tables of data that allow columns in one table to be used meaningfully in another table. If you have opened the Power BI Desktop example file PrestigeCarsRawSource.pbix, then you can see that there is a table of Makes and a table of Models. These two tables have been designed using a technique called relational modeling. Essentially this means that two tables have been created to avoid pointless data duplication. So any data that is used to describe a Make is stored in the Make table, whereas all the details concerning the models of vehicles are held in a separate table named Model. The two tables share a field that allows them to be joined so that users can see the data from the two tables together if they need to. In this specific example, the field that is common to both tables is MakeID.

Of course it is possible to store the data from these two tables as one table. However this would mean repeating elements such as the make name or make country each time that an invoice contained more than one item. Many IT systems prefer to avoid such duplication, and so store different levels and attributes of data in separate tables. However, this means that the two tables have to be linked together in the Power BI data model.

You can see another example of linking tables if you take a look (still in the PrestigeCarsRawSource.pbix file) at the Customer and Country tables. The Customer table contains a country column, but this column contains a country code, not the actual country itself. As a complement to this, there is a reference table of countries. It follows that, if we want to say what is the actual country name for a customer, we need to be able to link the tables so that the Customer table can look up the actual country of the customer. This requires some commonality between the two tables. Fortunately, the Country table contains a field CountryISO2 that contains the same codes that are present in the Country field of the Customer table. So if we are able to join the two tables using these fields, then we can see which country each customer resides in.

Clearly these examples are extremely simple. However, they are not unduly contrived, and represent the way many data sources store data. So there is every chance that you will see potential links, or relationships, like this in the real-world data that you will import from corporate databases or other sources. In any case, if you want to use data from multiple sources in your data analysis, you will have to find a way to link the tables using a common field. The reality may be messier (the fields may not have the same name in the two tables, for instance, and you may have to dig into the data to discover the fields that you can use to link tables), but the same core principle will always apply.

Creating Relationships

Creating relationships is easy once you know which fields are common between tables. Since you already saw that we need to join the Make table to the Model table using the MakeID field that both tables share, let's see how to do this.

1. Open the sample file PrestigeCarsRawSource.pbix.

2. Switch to the Model view.

3. Drag the MakeID field from the Make table over the MakeID field in the Model table as shown in Figure 1-4.

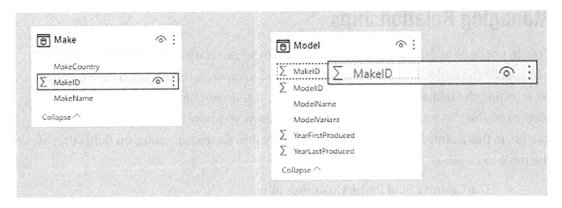

Figure 1-4. *Creating a table relationship*

This is all that you have to do. The two tables are now joined, and the data from both tables can be used meaningfully in reports and dashboards. Once a relationship is created, it will look something like Figure 1-5. The line between the tables indicates that a relationship between the tables exists.

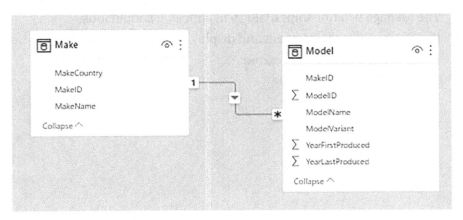

Figure 1-5. *Adding a relationship between two tables*

It is worth noting that there is no specific direction that you must follow when creating relationships. That is, you can drag the field you are joining the tables on from Make to Model or from Model to Make.

Note   Currently in the Power BI Desktop Data Model, you can only join tables on a single field.

Managing Relationships

You do not have to drag and drop field names to create relationships. If you prefer, you can specify the tables and fields that will be used to create a relationship between tables by selecting the tables and fields. What is more, you do not have to be in Relationship view to do this. So, just to make a point and to show you how flexible Power BI Desktop can be, in this example you will join the Country and Customer tables on fields that allow the table to be joined. These are

- The Country field in the Customer table

- The CountryISO2 field in the Country table

To create a relationship between the two tables

1. Switch to Report view.

2. In the Modeling ribbon, click the Manage Relationships button (this button is in the Home ribbon of the Data and Model views). The Manage Relationships dialog will appear. It should look like Figure 1-6 at the moment and display only the existing relationship that you just created.

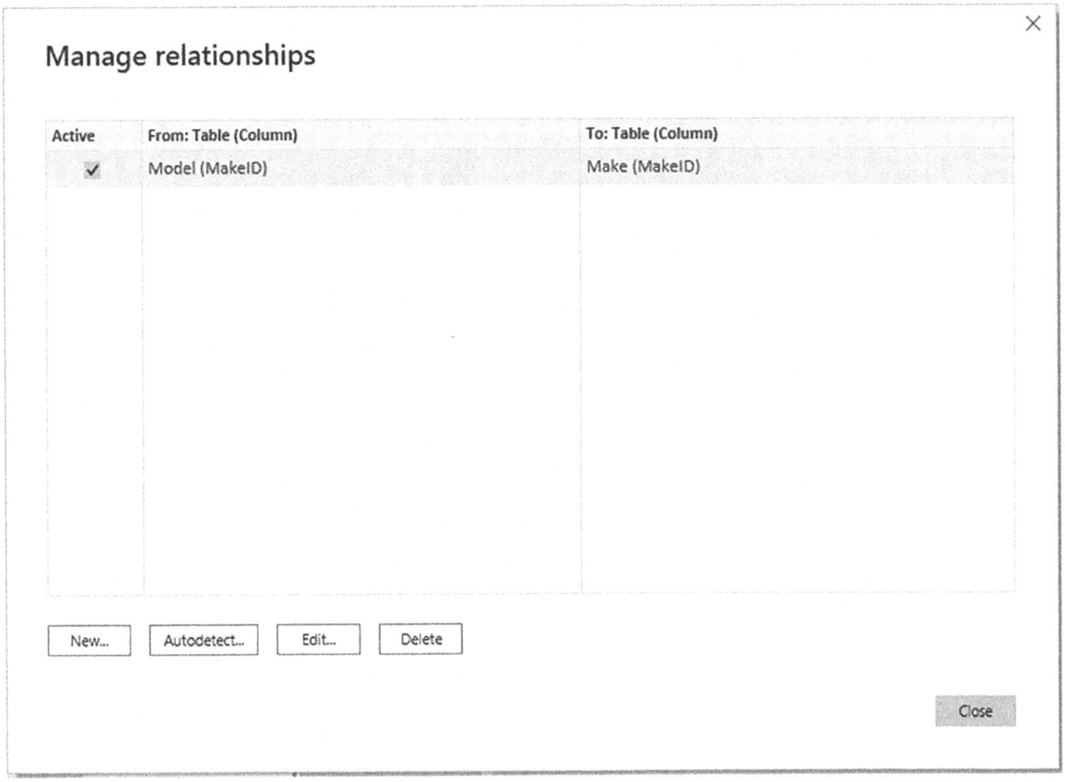

Figure 1-6. *The Manage Relationships dialog*

3. Click New. The Create Relationship dialog will appear

4. In the upper part of the dialog, select the Country table from the popup list of tables.

5. Click on the CountryISO2 field from the Countries fields.

6. On the lower row, select the Customer table as the Table you want to join the Country table to. If Power BI Desktop has guessed the field, it will appear selected. If it does not, or if it has guessed incorrectly, you can always select the correct field in the lower part of the dialog. In this example you will have to select the Country column. The Create Relationship dialog should look like Figure 1-7.

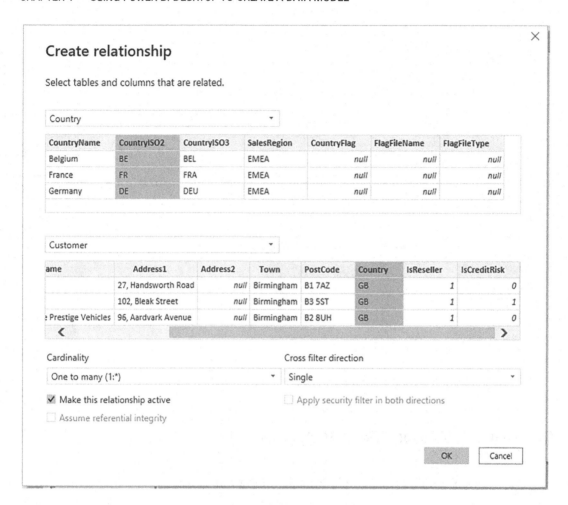

Figure 1-7. *The Create Relationship dialog*

7. Click OK. The Create Relationship dialog will close and return you to the Manage Relationships dialog.

8. Click Close. The Manage Relationships dialog will close, and the relationship will be created. The relationship will appear as a line joining the two tables in Model view.

Note   Creating relationships manually can be tougher when there are hundreds of fields in the source tables. If there are dozens of fields, you will have no choice but to scroll right through the tables to find the field that you need.

Examining Relationships

Once you have created one or more relationships, Model view displays the tables that are connected. However you have no way of seeing exactly which fields form the basis for the relationship. Fortunately, there is a simple solution to this challenge.

1. Hover the pointer over a relationship. The fields that make up the join in the two tables will be highlighted as shown in Figure 1-8.

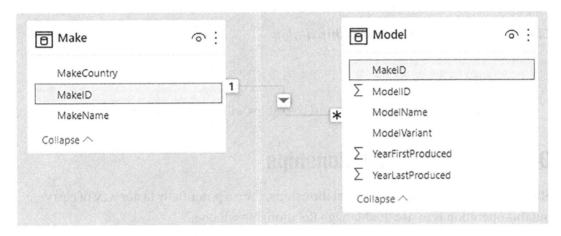

Figure 1-8. *Examining relationships*

Deleting Relationships

In addition to creating relationships, you will inevitably want to remove them at some point - especially if Power BI has autodetected incorrect relationships. This is both visual and intuitive.

1. Click on the Design View button in the Home ribbon. Power BI Desktop will display the tables in Relationship view.

2. Select the relationship that you want to delete. The arrow joining the two tables will become a double link, and the two tables will be highlighted.

3. Right-click and choose Delete (or press the Delete key). The Confirmation dialog will appear as shown in Figure 1-9.

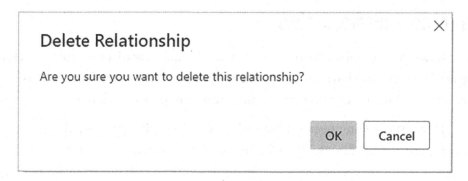

Figure 1-9. *The Delete Relationship dialog*

   4.   Click Delete.

The relationship will be deleted, and the tables will remain in the data model.

Deleting Multiple Relationships

Should you need to delete several relationships, then a potentially faster way of carrying out this operation is to use the Manage Relationships dialog.

   1.   Click on the Manage Relationships button in the Home ribbon.
        Power BI Desktop will display the Manage Relationships dialog.

   2.   Ctrl-click on all the relationships that you with to remove from the
        list of available relationships.

   3.   Click Delete. The confirmation dialog will appear.

   4.   Click Delete.

Note   If you delete a set of related tables and subsequently reimport them without importing the relationships, then Power BI Desktop will **not** remember the relationships that existed previously. Consequently, you will have to re-create any relationships manually. The same is true if you delete and reimport any table that you linked to an existing table in Power BI Desktop – once a relationship has been removed through the process of deleting a relationship or a table, you will have to re-create it.

Editing Relationships

As you progress in data modeling in Power BI Desktop, you may need to make subtle modifications to an existing relationship. This could involve switching the column used in one or both tables, for instance. If you wish to change the field in a table that serves as the basis for a relationship, then you can use the Manage Relationships dialog that you saw earlier.

1. In the Model window right-click the relationship that you wish to modify. A small popup menu will appear as shown in Figure 1-10.

Figure 1-10. *Modifying a single relationship*

2. Select Properties. The Edit relationship dialog will appear.

3. Click on the new column for either or both of the tables that you wish to use as the basis for the table join.

4. Click OK.

The relationship will now be based on the new column(s) that you have specified (assuming that you have specified columns that can be joined in a valid relationship).

The techniques used to create and manage relationships are not, in themselves, very difficult to apply. It is nonetheless *absolutely fundamental* to establish the correct relationships between the tables in the dataset. Put simply, if you try and use data from unconnected tables in a single Power BI Desktop dashboard, you will not just get an alert warning you that relationships need to be created, you will also get visibly inaccurate results. Basically, all your analysis will be false. So it is well worth it to spend some time upfront designing a clean, accurate, and logically coherent data model.

Now that you know how to join tables, you can practice by linking all the source tables in the PrestigeCarsRawSource.pbix file. The joins required are

- Make table to Model table on the MakeID field

- Model table to Stock table on the ModelID field

- Stock table to SalesDetails table on the StockCode field in the Stock table to the StockID field in the SalesDetails table

- SalesDetails table to Sales table on the SalesID field

- Sales table to Customer table on the CustomerID field

- Customer table to Country table on the Country field in the Customer table to the CountryISO2 field in the Country table

This model should look like the one shown in Figure 1-14.

Creating Relationships Automatically

Creating relationships is not difficult, providing that the columns in the tables to be linked exist. So, to save you time and facilitate the creation of data models, Power BI Desktop can attempt to join tables automatically when tables are loaded. Indeed, it will attempt to do this by default unless you ask it not to.

There are several points that are worth noting as far as this technique is concerned:

- You can avoid a lot of manual work - but the process can create superfluous relationships that you later have to delete manually.

- Power BI Desktop will try and create relationships between tables whatever the data source. This means that, for instance, a table based on a CSV file can join to another table sourced from Excel.

- Power BI Desktop will only attempt to create relationships where

  - The field names in the two tables are identical

  - The data types of the two fields in the two tables are identical

To see this in action you can try the following process:

1. Open a new, blank instance of Power BI Desktop.

2. Click File=>Options and Settings=>Options.

3. In the Current File section, click Data Load in the left pane.

4. Ensure that Autodetect new relationships after data is loaded is checked (this is the default).

5. Click OK to exit the Options and settings.

6. Click Excel Workbook in the Home menu.

7. Open the sample file PrestigeCars.xlsx in the sample data folder.

8. Select all the available worksheets.

9. Click Load.

10. Click the Model icon at the top left of the Power BI Desktop window.

11. Click Fit to Screen icon on the bottom right of the Power BI Desktop window. The data model should look like the one shown in Figure 1-11.

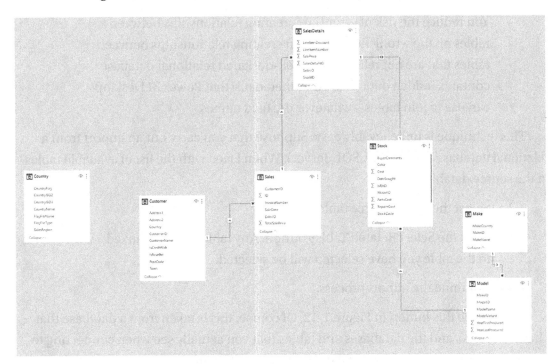

Figure 1-11. *Automated relationship creation*

You can see that all tables that share a common field with another table are linked. The only unlinked tables are Country and Customer – because the "shared" field does *not* have the same name in both tables.

Clearly creating relationships automatically like this can save a vast amount of time. However, you need to bear in mind that you will nonetheless need to ensure that the relationships are logically valid as Power BI Desktop is only using field names to link tables. Relationships created like this might not all be perfectly valid. So you should check that they are correct before creating and delivering dashboards.

Equally, if you are importing tables from a relational database, then you can request that Power BI Desktop to create the relationships automatically during the import process. This approach has a few advantages:

- You avoid a lot of manual work.

- Power BI Desktop will load required tables that you may not even have selected (or thought of).

- You reduce the risk of error (i.e., creating relationships between tables on the wrong fields, or even creating relationships between tables that are not related) as a well-designed relational database contains hidden data (known as metadata) that Power BI Desktop can use to join tables – whatever the field names.

This technique is unbelievably easy. Suppose that you carry out an import from a relational database source (say, SQL Server). When faced with the list of available tables in the source database, you

1. Select the major source table.

2. Click the Select Related Tables button. Any tables that are linked to the table you have selected will be selected.

3. Continue the import process.

You can see this button in Figure 1-12. Of course, this is taken from a database that I have access to, and the databases and tables that you actually see when connecting to your own SQL Server databases will be completely different.

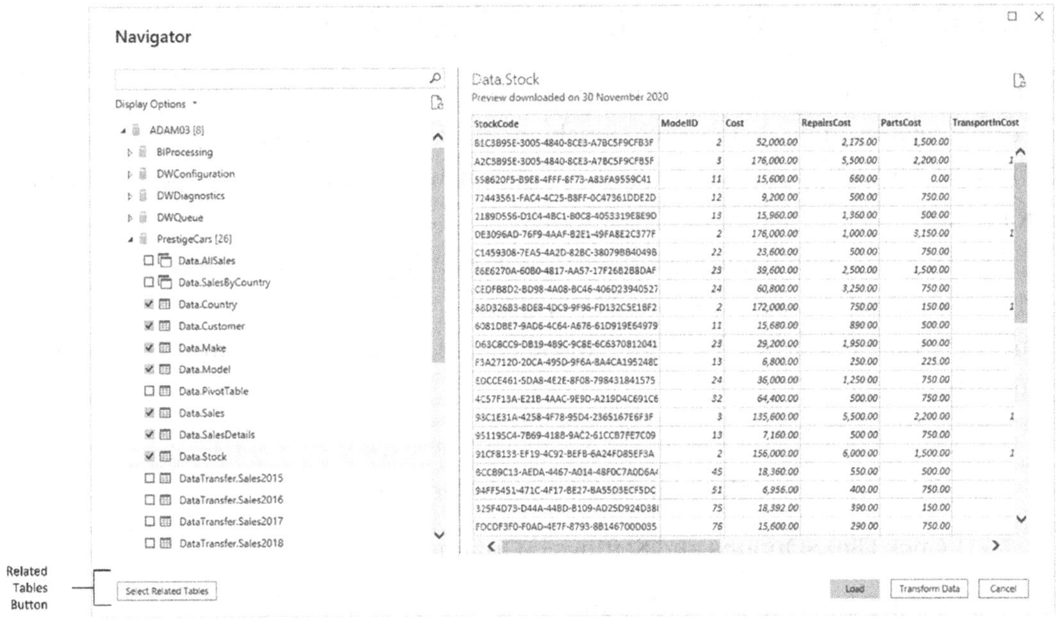

Figure 1-12. *Selecting related tables*

Once you have completed the import, switch to Relationship view. You will see that the tables you just imported already have the relationships generated in Power BI Desktop. This is particularly useful as relational databases often have many lookup tables that are required for analytics. This approach ensures that such tables are available in the data model.

Note   If you choose to select related tables, be aware that doing so only selects tables linked to the table(s) that you have already selected. As a result, you may have to carry out this operation several times, choosing a different starting table every time, to force all the existing relationships to be imported correctly.

Tip   If you have the necessary permissions as well as the SQL knowledge, then you can, of course, join tables directly in the source database using a query. This way you can create fewer "flattened" tables in Power BI Desktop from the start.

Inhibiting Relationship Autodetection

The core of the data model is the way tables are joined together. The joins between tables are called relationships. As table relationships are so fundamental to analytics, Power BI Desktop will always attempt to discover relationships between tables and join them automatically.

There may be times when you do *not* want Power BI Desktop to behave in this way. Perhaps the autodetection is unsuited to a particular set of source tables that you are using. Or (as is the case here) you want to look at the source data and apply any joins yourself manually.

To turn off relationship autodetection

1. Open a new, blank instance of Power BI Desktop.

2. Click File=>Options and Settings=>Options.

3. Click Data Load in the left pane (in the Current File section).

4. Uncheck "Autodetect new relationships after data is loaded."

The Options dialog should look something like the one shown in Figure 1-13.

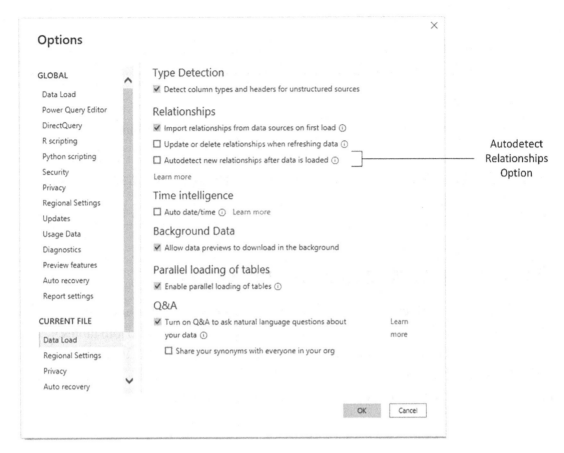

Figure 1-13. *The Options dialog*

5.   Click OK.

The one thing to stress at this point is that you must disable automatic detection of relationships before you load (or connect to) the source data that you will be using as the basis for your data model.

Deactivating Relationships

If you no longer need a relationship between tables but do not want to delete it, you also have the option of deactivating the relationship. This means that the relationship will no longer function, but that you can reactivate it quickly and easily should you ever need it later. This is explained in greater detail in Chapter 17.

To deactivate a relationship

1. In the Model tab, right-click on the required Relationship and select Properties. The Edit Relationship dialog appears.

2. Uncheck "Make this relationship active."

A deactivated relationship appears as a dotted line joining two tables.

Note   Be warned that queries across tables may no longer work correctly if you have deactivated a relationship.

The Meaning Behind Table Relationships

As you have seen in this chapter, creating a data model is technically quite simple. All you have to do is to tell Power BI Desktop which fields it must use to join tables. Once you have done this, you have a basic data model. If you open the Power BI file PrestigeCarsRelational.pbix and switch to the Model window, you will see a set of tables fully joined up into a basic data model. You can see this in Figure 1-14.

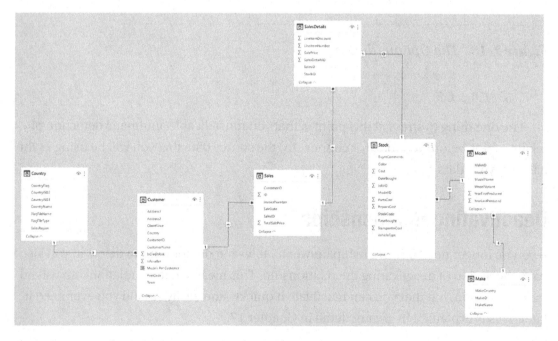

Figure 1-14. *The PrestigeCars Data Model*

The reason for spending time setting up the table joins is that now you can

- Use all the fields in all the joined tables in an interconnected way in Power BI visuals - almost as if you were using one vast table.

- Filter data in one table using fields in another table (in many - though not all - cases).

- Create DAX formulas that use data from multiple tables (as you will see starting in Chapter 5).

The data model does not make everything automatic, and there are constraints and limitations that you will learn about in upcoming chapters. Yet a joined up data model is the essence of analytics in Power BI. Without a coherent and structured data model, dashboard creation will be much, much more difficult. Not only that, but the analytical output could be catastrophically wrong.

Cardinality in Relationships

The Edit Relationship dialog contains one advanced option that it is essential to understand. Specifically this is the relationship's *cardinality*. This is a term that database designers and data modelers use to describe how the quantitative relationship of the contents of one field relate to the contents of another field. That is

- Is there only one example of each piece of data in one table yet many examples of the same data in the other table's column? This is the case for "lookup" tables where both the code and the description of what a code means are stored once in one lookup table but the code is used in other tables many times.

- Is a piece of data used many times in one table - a detail table - yet only found once in another table? This can be the case in "Parent-child" tables. An example of this is an invoice header table that contains invoice details such as the invoice number, date and client information, and an invoice lines table that contains a row for each product sold.

You can see the available cardinality options in Figure 1-15.

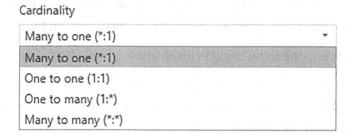

Figure 1-15. *Relationship cardinality*

These various options are described in Table 1-3.

Table 1-3. *Power BI Desktop Relationship Cardinality*

Relationship Option	Description
Many-to-One	Specifies that there are many records in one table for a single record that maps in the table that is joined
One-to-One	Specifies that there is a single record in one table for a single record that maps in the table that is joined
One-to-Many	Specifies that there is a single record in one table for many records that map in the table that is joined
Many to Many	Specifies that there are many records in one table for many records that map in the table that is joined

Cardinality Indicators

The Model window in Power BI Desktop indicates what the cardinality of each join is. If you look back to Figure 1-8, you can see that the Make table has a 1 where the join to the Model table starts and the Model table has an asterisk (*) where the join to the Make table starts. Table 1-4 explains these cardinality indicators.

Table 1-4. *Power BI Desktop Cardinality Indicators*

Indicator	Type	Description
1	One	Indicates that there is only one unique example of each data element in the table. That is, there are no duplicates
*	Many	Indicates that a data element may be repeated on multiple rows of the table

Conclusion

This chapter was all about taking the clean data that you had prepared using Power Query in Power BI Desktop and molding it into a structured and coherent data model that will be the basis for your dashboards and reports. Should you need to learn all about Power Query, then the companion volume "Pro Data Mashup for Power BI" should answer your questions about this aspect of Power BI Desktop.

You learned the techniques required to link (or join if you prefer) disparate data sources together in a structured data model that can become the basis for in-depth analytics.

Now you need to move on to a deeper look at the data model in Power BI and the approaches and techniques required to apply conformity to the data itself. This is what you will see in the next chapter.

CHAPTER 2

Extending the Data Model

In the first chapter you learned how to join tables to create a data model that you can use to analyze data across a set of tables. This is the necessary first step to delivering a structured data model. However, it is only a first step. This initial structure needs to be developed and polished to enable it to deliver the analytics that you require. So in this chapter you will see how to extend the core data model by

- (Re)naming tables and fields

- Specifying data types

- Categorizing data

- Formatting data in the data model

- Adding "sort by" fields that ensure the correct sort order in dashboard elements

- Defining key fields

Once again, all the sample files used in this chapter are available on the Apress website. Once downloaded they should be in the folder C:\PowerBiDesktopSamples.

The Properties Pane

In this chapter we will be using the Properties pane in the Model view. You need to be aware that the Properties pane contents vary according to whether you have selected

- The model canvas

- A table

- A field

The three properties panes can be seen in Figure 2-1.

© Adam Aspin 2023
A. Aspin, *Pro DAX and Data Modeling in Power BI*, https://doi.org/10.1007/978-1-4842-8995-2_2

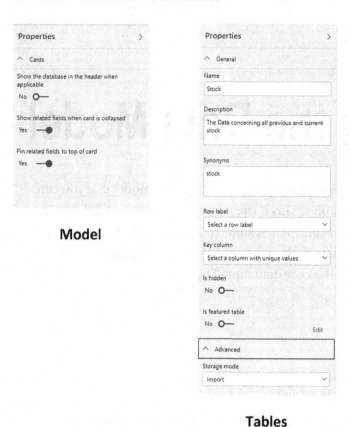

Model

Tables

Fields

Figure 2-1. *The properties panes*

Note   The Properties Pane can be expanded and collapsed like any other Power BI Desktop pane. Simply click the chevrons at the top right of the Properties pane to display or hide it.

Managing Power BI Desktop Data

Assuming that all has gone well, you now have a series of tables from various sources successfully added to your Power BI Desktop data model. You can see these data tables in the Fields List at the right of the Power BI Desktop window. Clicking on a table will display the data from that table in the central area of the Data View window. It will soon be time to see what we can do with this data, but first, to complete the roundup of overall data model management, you need to know how to do a few things like

- Rename tables

- Delete tables

- Move tables

- Move around a table

- Rename a field

- Delete fields

Manipulating Tables

Let's begin by seeing how you can tweak the tables that you have imported. To try out these techniques, you can load the sample file PrestigeCarsRelational.pbix from the sample data folder.

The data model currently looks like the one that you saw at the end of the previous chapter in Figure 1-14. You can see this by switching to the Model view.

Renaming Tables

Suppose that you wish to rename a table that you previously imported using Power Query. These are the steps to follow:

1. Click on the table header in the model canvas (be sure not to click on a field name).

2. Change the name in the Name box in the Properties pane.

Alternatively (and in any Power BI Desktop view), you can

1. Right-click on the table name in the Fields pane (or click on the ellipses to the right of the table name).

2. Select Rename. The current name will be highlighted in the fields list.

3. Enter the new name, or modify the existing name.

4. Press Enter.

Alternatively you can double-click on the field header, or press F2 when the field header is selected.

This will also rename the query on which the table is based. In essence Power BI will let you rename datasets either as queries in the Power Query or in the Fields Pane. As the query *is* the table, renaming one renames the other.

Deleting a Table

Deleting a table in Model view can be done by

1. Right-click on the table name in the Model canvas (or click on the ellipses to the right of the table name). The popup menu will appear, as shown in Figure 2-2.

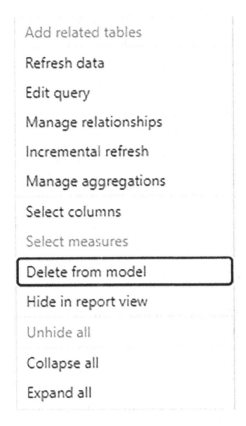

Figure 2-2. *The Model canvas popup menu*

2. Select Delete from model. The delete table confirmation dialog
    will appear as shown in Figure 2-3.

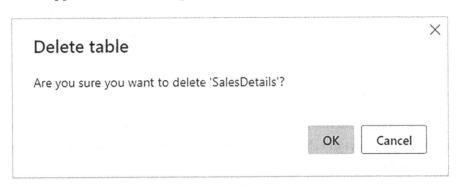

Figure 2-3. *The delete table confirmation dialog*

3. Click OK to remove the table from the data model.

You can also delete a table from the Fields pane by right-clicking on the table name (or clicking the ellipses to the right of the table name) and selecting Delete from model from the context menu. You can also just press "Delete" on your keyboard.

Note   When you delete a table, you are removing it from Power BI Desktop *completely*. This means that it will also be removed from the set of queries in Power Query that you may have used to transform the data. So you need to be careful because you could lose all your carefully wrought transformation steps as well.

Manipulating Fields

Now let's see how to perform similar actions — but this time inside a table — to the fields of data that make up the table.

Renaming a Field

Renaming a field is pretty straightforward. All you have to do is

1. Click on the field name in the model canvas.

2. Change the name in the Name box in the Properties pane.

Alternatively (and in any Power BI Desktop view) you can

1. In the Field List, right-click on the field that you wish to rename (or click on the ellipses to the right of the field name).

2. Select Rename from the context menu.

3. Type in the new name for the field and press Enter.

And that is it; your field has been renamed. You cannot, however, use the name of an existing field in the same table. Similar to renaming tables, you can double-click or press F2 to rename a field.

Note   Power BI Desktop is very forgiving when it comes to renaming fields (or indeed tables and calculations too). You can rename most elements in the Query Window, the Data View, the Report View, or the Relationships View (depending where they were added) and the changes will ripple through the entire Power BI Desktop data model. Better still, renaming fields, calculations, and tables will generally not cause Power BI Desktop any difficulties if these elements have already been applied to dashboards, as the dashboards will also be updated to reflect the change of name.

Deleting Fields

Deleting a field is equally easy. You will probably find yourself doing this when you have either brought in a field that you did not mean to import, or you find that you no longer need a field. So, to delete a field

1. Right-click on the table name in the Model canvas (or click on the ellipses to the right of the table name). The popup menu will appear, as shown in Figure 2-4.

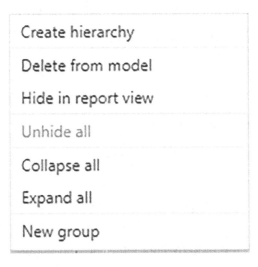

Figure 2-4. *The field context menu*

2. Select Delete from model. The delete column confirmation dialog
   will appear as shown in Figure 2-5.

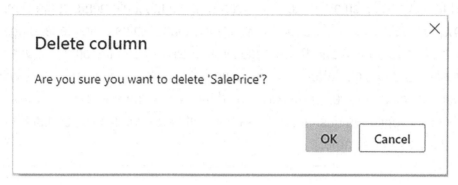

Figure 2-5. *The delete field confirmation dialog*

3. Click OK to remove the field from the data model.

You can also delete fields from the Fields pane in any of the Power BI Desktop views.
This is done by

1. In the Field List, right-click on the field that you wish to delete
   canvas (or click on the ellipses to the right of the field name).

2. Select Delete from model from the context menu (or press Delete).

3. Click OK in the confirmation dialog.

Deleting unused fields is good practice, as this way you will

- Reduce the memory required for the dataset.

- Speed up data refresh operations.

- Reduce the size of the Power BI Desktop file.

Note   Deleting a field really is permanent. You cannot use the undo function to
recover it. Indeed, refreshing the data will not add the field back into the table
either. If you have deleted a field by accident, you can choose to close the Power
BI Desktop file without saving and reopen it, thus reverting to the previous version.
Otherwise you can open Power Query and add the field name back into the query
from the original data source.

Moving Fields

Once in the Data Model you *cannot* move fields around. So if you want to change the field order for any reason, you will have to switch to Power Query and move the field there. Once you save and apply your changes, the modified field order will be visible in the Data Model. This will not affect the order of fields in the Fields pane - these are always in alphabetical order.

Power BI Data Types

When you are importing data from an external source, Power BI Desktop will try and convert it to one of the nine data types that it uses. These data types are described in Table 2-1.

Table 2-1. *Power BI Desktop Data Types*

Data Type	Description
Decimal number	Stores the data as a real number with a maximum of 15 significant decimal digits. Negative values range from -1.79E +308 to -2.23E -308. Positive values range from 2.23E -308 to 1.79E + 308
Fixed decimal number	Stores the data as a number with a specified number of decimals
Whole number	Stores the data as integers that can be positive or negative but are whole numbers between 9,223,372,036,854,775,808 (-2^63+1) and 9,223,372,036,854,775,807 (2^63-2)
Date/time	Stores the data as a date and time in the format of the host computer. Only dates on or after the 1st of January 1900 are valid
Date	Stores the data as a date in the format of the host computer
Time	Stores the data as a time element in the format of the host computer
Text	Stores the data as a Unicode string of 536,870,912 bytes (268,435,456 characters) at most
True/false	Stores the data as Boolean – true or false
Binary	Stores the data as binary (machine-readable) data

You need to be aware that you can only change a data type if *all* the data in the field (or column if you prefer) can be interpreted as the data type that you select. For instance, if you define a column as Whole Number when the column contains some records containing text in this field, then the data type change will not be allowed.

So why would you need to change data types? Well, some potential reasons could be

- You are looking at a field that contains numbers that will never be used for calculations. Telephone numbers or certain postcodes are examples of these. It is good practice to set fields like these to text.

- You wish to truncate decimals for certain numeric fields as you will never need the decimal accuracy – and wish to avoid rounding errors.

- You wish to reduce the size of the data model by excluding decimals or time parts of dates.

- Power BI has presumed that certain fields contain text elements – whereas you know them to be dates. Converting to a date data type allows you to use dates in calculations more easily (as you will see in Chapters 7, 13, and 14).

Formatting Power BI Desktop Data

Power BI Desktop allows you to apply formatting to the data that it contains. When you format the data in the Power BI data model, you are defining the format that will be initially used in visualizations in all the dashboard pages that you create using this metric. So it is probably worth learning to format data for the following reasons:

- You will save time and multiple repetitive operations when creating reports and presentations by defining a format once and for all in Data View. The data will then appear using the format that you applied in multiple visualizations in this Power BI Desktop file.

- It can help you understand your data more intuitively if you can see the figures in a format that has intrinsic meaning.

Here is how to format a field (of numbers in this example):

1. Assuming that you are working in the Power BI Desktop Model view, click on the field that you want to format (TotalSalePrice from the Sales table in the sample file PrestigeCarsRelational.pbix, for instance, if you are using the sample data).

2. In the Properties pane, click on the Format popup. This is shown in Figure 2-6.

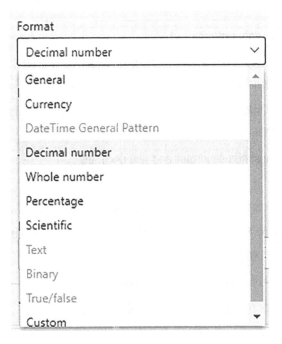

Figure 2-6. *The Format popup in the Properties pane*

3. Select Decimal number.

4. Set Thousands separator to On.

5. Set Decimal places to 2. All the figures in the field will be formatted with a thousands separator and two decimals.

Formatting is such a fundamental aspect of a data model that it can be applied in several different places. Alternatives to the techniques that you just saw are

1. In Data view, select the table and then the column that you want to format, and use the formatting options available in the Column tools menu.

2. In Report view select the table and then the column that you
    want to format in the Fields pane, and use the formatting options
    available in the Column tools menu.

The various formatting options available are described in Table 2-2.

Table 2-2. *Currency Format Options*

Format Option	Description	Example
General	Leaves the data unformatted	100000.01
Currency	Adds a thousands separator and two decimals as well as the current monetary symbol	£100,000.01
Date/time	Formats a date and/or time value in one of a selection of date and time formats	
Decimal number	Adds a thousands separator and two decimals	100,000.01
Whole number	Adds a thousands separator and truncates any decimals	100,000
Percentage	Multiplies by 100, adds two decimals, and prefixes with the percentage symbol	28.78%
Scientific	Displays the numbers in scientific format	1.00E+05
Decimal point	Increases or reduces the number of decimals	
Custom	Allows you to define custom formats (these are explained in Chapter 3)	

If you wish to return to "plain vanilla" data, then you can do this by selecting the
General format. Remember that you are not in Excel, and you cannot format only a range
of figures — it is the whole field or nothing.

Fortunately you *can* format nonadjacent fields by Ctrl-clicking to perform a
noncontiguous selection even across multiple tables in the Model canvas.

By now you have probably realized that Power Bi operates on a "Format once-apply
everywhere" principle. However this does not mean that you have to prepare the data
exhaustively before creating dashboards and reports. You can flip between the data
model and the Report view at any time to select another format; secure in the knowledge
that the format that you just applied will be used throughout all your dashboards
wherever the relevant metric is used (unless you override it on a particular visual).

Note   Numeric formats are not available for selection if the data in a field is of text or data/time data type. Similarly date and time formats are only available if the field contains data that can be interpreted as dates or times.

Preparing Data for Dashboards

Corralling data into a structure that can power your dashboards necessitates a good few tweaks above and beyond specifying data types and formats for final presentation. As part of the groundwork for your dashboards, you could also have to

- Categorize Data

- Apply a default summarization

- Define Sort By fields

- Set a key column

These ideas probably seem somewhat abstruse at first sight - so let's see them in action to make it clear why you need to add these touches to your data model.

Categorize Data

Power BI dashboards are not just made up of facts and figures. They can also contain geographical data or hyperlinks to websites or documents. While we humans can recognize a URL pretty easily, and can guess that a field with postcodes contains, well, postcodes, such intuitions may not be quite as self-evident for a computer.

So if you want Power BI to be able to add maps or hyperlinks (for instance), you will make life easier for both you and the application if you categorize any fields that contain the types of data that are used for maps and links.

For instance, suppose that you want to prepare the Country table a potential data source for a dashboard map (and assuming that you have loaded the sample file PrestigeCarsRelational.pbix).

1. In the Model view click inside the CountryName field in the Country table. The field will be highlighted.

2. In the Properties pane expand the Advanced section.

3. Click on the Data category popup. The available options will be displayed as shown in Figure 2-7.

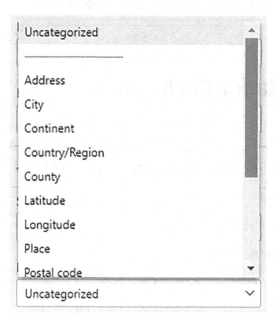

Figure 2-7. *The data category popup in the Properties pane*

4. Select Country/Region from the menu.

This means that Power BI Desktop now knows to use the contents of this field as a country when generating maps in dashboards.

Note   You can also apply categorization from the Column tools menu that appears whenever you select a table and column in the Fields pane in either of the Report or Data views.

The data category options that are available are described in Table 2-3.

Table 2-3. *Data Category Options*

Data Category Option	Description
Uncategorized	Applies to data that will not be used for hyperlinks or creating maps
Address	Specifies an address for mapping
City	Specifies a city for mapping
Continent	Specifies a continent for mapping
Country/region	Specifies a country or region for mapping
County	Specifies a county for mapping
Latitude	Specifies a latitude for mapping
Longitude	Specifies a longitude for mapping
Place	Specifies a location or place for mapping
Postal code	Specifies a postal (zip) code for mapping
State or province	Specifies a state or province for mapping
Web URL	Indicates an URL for a hyperlink
Image URL	Indicates to Power BI Desktop that the image referenced in the URL is to be used (once the dashboard is published)
Barcode	Displays a barcode (once the dashboard is published)

Note   Not specifying a data category does not mean that Power BI Desktop cannot create maps in dashboards or recognize URLs. However the results cannot be guaranteed of a reasonable chance of success unless you have indicated to the application that a field contains a certain type of data.

Apply a Default Summarization

When you are creating dashboards, you will nearly always be aggregating numeric data. Most times this will mean adding up the figures to return the field total. As you have probably seen when creating dashboards, this happens automatically whenever you add a numeric field to any visual.

There could be some fields where you will want another aggregation – as opposed the default summation of the field - applied most times. To set your own default aggregation (assuming that you have loaded the sample file PrestigeCarsRelational.pbix)

1. In the Model view select the TotalSalePrice field in the Sales table.

2. In the Properties pane expand the Advanced section.

3. Click Summarize by popup. The popup list will appear as shown in Figure 2-8.

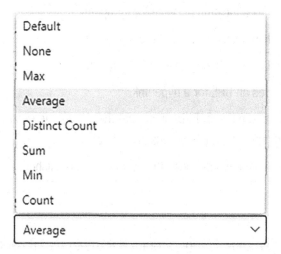

Figure 2-8. *The Summarize by popup*

4. Select Average from the menu.

Note   You can also apply a default summarization from the Column tools menu that appears whenever you select a table and column in the Fields pane in either of the Report or Data views.

The default summarization options that are available are described in Table 2-4.

Table 2-4. *Default Summarization Options*

Default Summarization Option	Description
Do not summarize	The data in this field will not be summarized
Sum	The data in this field will be added up (summed)
Average	The average value for the data in this field will be returned
Minimum	The minimum value for the data in this field will be returned
Maximum	The maximum value for the data in this field will be returned
Count	The number of elements in the field will be returned
Count (distinct)	The number of individual (distinct) elements in the field will be returned

Obviously you can only apply mathematical aggregations to numeric values. However you can apply counts to any type of data.

Note Specifying a default aggregation does not prevent you overriding the default in dashboards. It merely sets a default that will be applied as a standard when aggregating data from a field.

Define the Sort by Column

Sometimes you will want to sort data in a dashboard visualization based not on the contents of the selected field (or column if you prefer), but by the contents of another field (or column). As an example, imagine that you have a table of data that contains the month for a sale. If you sort by month, you probably do not want to see the months in alphabetical order, starting with April. In cases such as this you can tell Power BI Desktop that you want to sort the month *name* element by the month *number* that is contained in another field.

1. In the Model view select the field that you wish to sort using the contents of another field as the basis for the sort operation.

2. In the Properties pane expand the Advanced section.

3. Click Sort by popup. The popup list will display all the other fields in the table containing the field that you selected.

4. Click on the field to sort by.

Assuming that you had applied this technique to the names of months in a visual, if you now sort by a month name in a visual, you will see the months in the order that you probably expect - from January to December. Had you *not* applied a Sort By field then calendar months would have been sorted in alphabetical order - that is, from April to September! Once again this choice will apply to *any* visualization that you create in a Power BI Desktop dashboard that is based on this data model.

Real-world data may not always contain fields that you can use as Sort By fields. So remember that you can always switch to Power Query and enhance source tables with extra fields that you can then use to sort data in Data View.

Note   You can also apply a sort by column from the Column tools menu that appears whenever you select a table and column in the Fields pane in either of the Report or Data views.

Key Columns

Many databases and data models need a single element that can allow a row to be considered unique. In a lookup table (such as the Make table in the sample data model), this is the MakeID field that contains a *unique* number for each make of car.

It can help Power BI if you indicate what these fields are in the data model. This is because setting a field as a key forces Power BI to ensure that there are no null values or duplicates in these fields (otherwise an error message appears). To do this

1. In the Model view select the table where you want to define as a key.

2. In the Key column popup select the field in the table that represents a unique value for every row.

Nullable Fields

In the world of databases, it can be important to ensure that certain fields do not allow blanks – or NULLs as data people call them. This is because having empty elements in a column can, in certain cases, skew the analyses that use these fields.

Power BI allows you to specify that a field may not contain NULLs (which is also defined as setting its nullability).

1. In the Model view select the field whose nullability you wish to set.

2. In the Properties pane expand the Advanced section.

3. Set Is nullable to On.

Note   You can only define that a field is nullable if the current field does not contain any blanks.

Once a field is set not to allow NULLs, the data refresh will fail if the source data actually contains NULLs. You may wish to add a step in the Power Query process that loads tables like these to replace empty data with a recognized "unknown" value such as "N/A."

The Power BI Desktop Data View

Modeling data in Power BI Desktop does not only use the Model view. Many of the settings that you can configure in the Model view can also be set (or viewed) in the Data view.

To switch to Data view

1. Click on the Data icon on the top left of the Power BI Desktop window as shown in Figure 2-9.

Figure 2-9. *Data view*

Data view will display the data contained in the tables in the data model. To display the contents of a specific table, simply click on the table (or any field in the table) in the Fields pane.

One fundamental use of the Data view is to allow you to model the data based on an understanding of what the data is. Data view allows you to scroll down through a complete dataset – unlike queries in Power Query, where you only can see a sample of the source data.

There are two ribbons that are specific to the Data view:

- Table tools

- Column tools

These ribbons allow you to examine and change many of the attributes that you defined in the Properties pane of the Model view.

The Table Tools Ribbon

The table tools ribbon lets you carry out certain essential modifications to the selected table – or even add a new table for ad-hoc data as you will learn in Chapter 3.

You can see the buttons that are available in Figure 2-10.

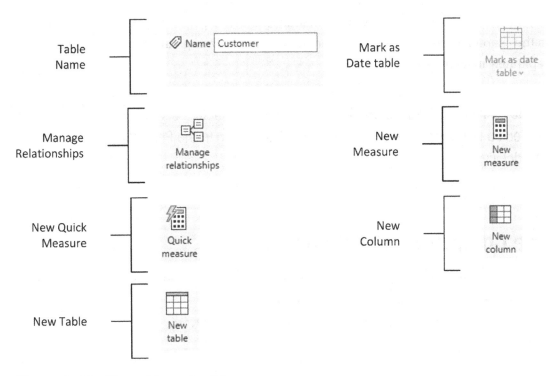

Figure 2-10. *The table tools ribbon*

The ribbon buttons are explained in Table 2-5.

Table 2-5. *Table Tools Buttons*

Button	Description
Name	Displays (and allows you to edit) the table name
Mark as date table	Sets the selected table as a date table that can be used with time intelligence functions
Manage relationships	Allows you to manage relationships (joins) between tables
New measure	Creates a new measure in the selected table
Quick measure	Displays the Quick Measures dialog to help you produce common DAX measures more quickly and easily
New column	Adds a new column to the selected table
New table	Creates a new manual table

As the table tools buttons Name and Manage relationships were explained earlier in the chapter, I will not re-explain them here. The remaining buttons are explained separately in later chapters.

Note   Any modifications that you make using the Table tools ribbon are reflected throughout the entire data model, including in the Properties pane of the Model view.

The Column Tools Ribbon

The column tools ribbon lets you carry out certain essential modifications to the column that you have clicked on inside the selected table.

You can see the buttons that are available in Figure 2-11.

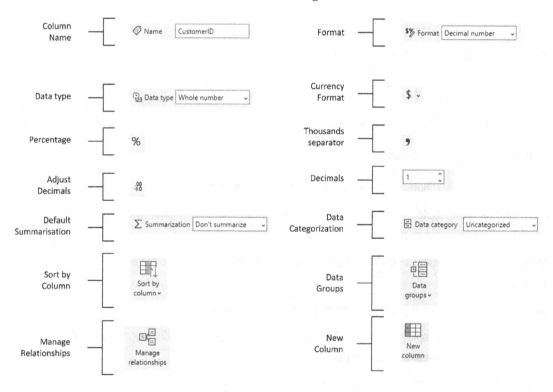

Figure 2-11. *The column tools ribbon*

The column tools ribbon buttons are explained in Table 2-6.

Table 2-6. *Column Tools Buttons*

Button	Description
Name	Displays (and allows you to edit) the column name
Data type	Displays (and allows you to edit) the data type of the data in the column
Format	Selects a numeric format for the data in the column
Currency format	Applies a currency format to the data in the column
Percentage format	Applies a percentage format to the data in the column
Comma-separated format	Applies a comma-separated format to the data in the column
Change decimals	Changes the number of decimals
Number of decimals	Modifies the number of decimals
Default summarization	Selects a default summarization
Data category	Applies a data category
Sort by column	Defines the column used to order the data for the selected column
Manage relationships	Allows you to manage relationships (joins) between tables
New column	Adds a new column to the selected table

As all functions applied using the column tools buttons (except data groups and new column) were explained earlier in the chapter, I will not re-explain them here. Data groups are explained in Chapter 3 and new columns in Chapter 4.

Note   Any modifications that you make using the Column tools ribbon are reflected throughout the entire data model, including in the Properties pane of the Model view.

Sorting Data in Power BI Desktop Tables

A Power BI Desktop table could contain millions of rows, so the last thing that you want to have to do is to scroll down through a random dataset. Fortunately ordering data in a table is simple:

1. Right-click inside the column header of the field you want to order the data by. The context menu appears (as shown in Figure 2-12).

Figure 2-12. *The Data view context menu*

2. Click the Sort Ascending option in the context menu to sort this field in ascending (alphabetical) order.

The table will be sorted using the selected field as the sort key, and even a large data set will appear correctly ordered in a very short time. If you want to sort a table in descending (reverse alphabetical or largest to smallest order) order, then click on the

Sort Descending option in the context menu to sort this field in descending (largest to smallest or reverse alphabetical) order.

You need to remember at this juncture that the Data Model is not really designed for interactive data analysis. That is what dashboards are for. Consequently you should not expect to use the data tables in Power BI Desktop as if they were vast Excel spreadsheets. Moreover, sorting the data in the data table will have no effect on the way that data appears in visuals as visuals are sorted individually.

Tip   If you need a visual indication that a field is sorted, look at the right of the field name. You will see a small arrow that faces downward to indicate a descending sort or upward to indicate an ascending sort.

If you want to remove the sort operation that you applied and return to the initial data set as it was imported, all you have to do is click the Clear Sort icon in the context menu for the field.

Note   You cannot perform complex sort operations; that is, you cannot sort first on one field, then – carry out a secondary sort in another field (if there are identical elements in the first field). You also cannot perform multiple sort operations sorting on the least important field and then progressing up to the most important field to sort on to get the effect of a complex sort. This is because Power BI Desktop always sorts the data based on the data set as it was initially loaded. Remember that you can add index fields in Power Query and then sort on these if you want to reapply an initial sort order. This is explained in the companion volume *Pro Data Mashup with Power BI*.

Setting Field Widths

One final thing that you may want to do in data view to make your data more readable — and consequently easier to understand — is to adjust the field width. I realize that as an Excel or Word user, you may find this old hat, but in the interests of completeness, here is how you do it:

Place the mouse pointer over the right-hand limit of the field title in the field whose width you want to alter. The cursor will become a two-headed arrow.

1. Drag the cursor left or right.

Note   You cannot select several adjacent fields before widening (or narrowing) one of them to set them all to the width of the field that you are adjusting. You can, however, double-click on the right-hand limit of the field title in the field whose width you want to alter to have Power BI Desktop set the width to that of the longest element in the field.

Currency Formats

Power BI Desktop will propose a wide range of currency formats. To choose the currency that you want

1. Click on the popup (the downward-facing triangle) to the right of the currency format icon. This will display a list of available formats.

2. Select the currency symbol that you want, or scroll through the list to view all the available currency formats, as shown in Figure 2-13, and then click OK.

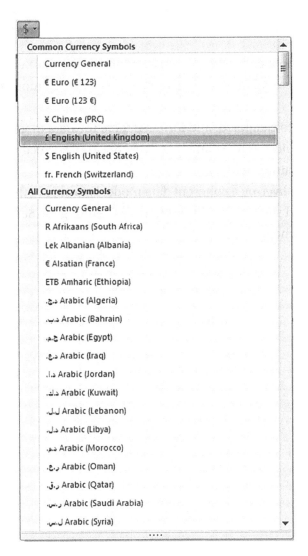

Figure 2-13. *The currency format popup list*

Note   The thousands separator that is applied, as well as the decimal separator, will depend on the settings of the PC on which the formatting is applied.

Conclusion

This chapter was all about enhancing a core data model. You learned how to remove, rename, and hide tables and fields as well as converting data to more appropriate data types.

Then you learned to categorize and format data in the data model. You then saw how to apply sort by columns to ensure that data appeared correctly in visuals as well as categorizing and formatting fields.

The final step in delivering a coherent data model is to ensure that it is easy to understand by you and your users. This is known as defining the Semantic Model and is described in the next chapter.

CHAPTER 3

The Semantic Layer

In the two previous chapters, you learned how to take separate tables and join them together in a way that allows you to query data across multiple source data elements. You then saw how to add further elements to shape the data model to make it easier to use and maintain.

Yet what you have seen so far only takes the data "as-is" and makes it possible to deliver analytics. The next stage in data modeling with Power BI Desktop is to make the data model

- Easier to use

- More intuitive and immediately comprehensible

- Better able to deliver results faster

These aims mean that the data model will be enhanced – and in many cases reworked – to add a *semantic layer*. Adding this layer to the data model creates a *semantic model*.

The driver behind the semantic layer is the aim of *representing corporate data in terms that business users understand immediately*. This means moving away from IT-provided data structures and terms towards a simplified model of the data that accelerates time to insight - and provides consistent results.

In many ways the semantic layer maps complex data into familiar business terms and also offers a unified and consolidated view of corporate data. It should also deliver a single version of the truth and allow users to analyze and report on the same underlying reality. Hopefully it can provide a common "look and feel" when accessing and analyzing data.

The semantic layer is, at its heart, the data model. After all, this is the structure that has the potential to hide the complexity of the underlying data sources and creates a simpler view of the source data.

© Adam Aspin 2023
A. Aspin, *Pro DAX and Data Modeling in Power BI*, https://doi.org/10.1007/978-1-4842-8995-2_3

However, a semantic model can be much more than this. It can and should entail

- A representation of the data in the way that business – not IT – conceive the organization

- Clear and comprehensible names for tables and fields

- Descriptions or even explanations of terms and calculations

- A naming convention that is standardized and guides users

- Only required fields being visible, and all superfluous data excluded

- Data organized in a clear, comprehensible, and accessible way

I cannot deny that creating a good semantic layer can be a lot of work. However, the benefits can, equally, be considerable. The good news is that you have already learned many of the techniques that are required to add the main elements of the semantic layer. So what we will be looking at in this chapter is *how* these techniques can be applied to push the data model to the next level and make it easy – and even fun – for users to rely on to create their dashboards in Power BI Desktop.

In this chapter we will look at

- Topologies

- Field and table names

- Descriptions

- Hiding tables and fields

- Hierarchies

- Folders

- Binning and grouping

- Groups

- Naming convention

It is worth remembering that the semantic layer is, essentially, a reworking and extension of the underlying structures that you see in the Data and Model views – and that you use in the Report view.

Data Model Topologies

The first step in adding greater clarity to a data model is probably far and away the most difficult. This is because many initial data structures – such as the one that you have seen so far in this book – are not optimized for simplicity and clarity.

It is possible in most cases to restructure the source data so that the data model is based around

- *Facts* – which are the numbers that will be aggregated to deliver meaningful analysis

- *Dimensions* – which are the descriptive elements that will be used to structure and filter the source data

A model like this (which is known as a dimensional model) can be based on the available source data, yet can be considerably easier to use.

This model is probably best appreciated if you load the sample file PrestigeCarsDimensional.pbix where the tables and data have been reorganized into a dimensional model. If you look at Figure 3-1 which shows the Model pane from this file you can see the seven tables that you met in the two previous chapters restructured into

- *One fact table* – containing all the metrics required for analysis plus any "link" fields needed to join tables

- *Four dimensions* – where the descriptive elements needed to structure and filter the source data have been created according to a user-friendly structure

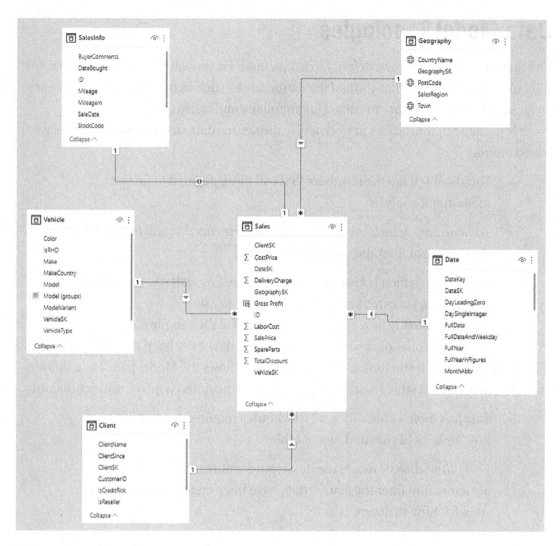

Figure 3-1. *Star schema topology*

Note   This data model contains a date dimension that helps you present and calculate data over time. Creating a date dimension is explained in Chapter 13.

Creating a dimensional model can take weeks (and possibly require months or even years) of experience. So I will not attempt to condense or resume the hundreds of books and articles that exist on the subject. However I would like to underline that a business-orientated and user-friendly model like this can make analytics considerably easier. Indeed, I will be using this model for the rest of the book as it simplifies the DAX that you will be learning from the next chapter onwards.

Note   If you want an idea of some of the Power Query techniques that you can use to adapt a relational data model to a dimensional data model, a good starting point is my article "Power BI For Data Modelling" on *Simple Talk*. You can find this at `www.red-gate.com/simple-talk/databases/sql-server/bi-sql-server/power-bi-data-modelling`.

I fully understand that many IT professionals simply do not like dimensional models. So I am not suggesting that you *have* to use this topology. However, I really do believe that it can suit many reporting model requirements in Power BI. Also it is, in my experience, the optimal model structure for Power BI.

Field and Table Names

Although renaming a field may seem trivial, it can be an important factor in making a data model usable. Consider other users first; they need fields to have instantly understandable names that mean something to them. Then there is the Power BI Q&A natural language feature. This will only work well if your fields have the sort of names that are used in the queries — or ones that are recognizable synonyms.

So, as simple as it may be, giving the fields in the data model instantly clear and comprehensible names can turbocharge both acceptance of the data model by your users and the speed with which you and they develop powerful dashboards.

Descriptions

While modifying field names to make them comprehensible is a major step on the path to a clear semantic layer, adding extended descriptions to tables and fields can help users to understand the data model even better.

To add a description to a table or field

1. Switch to Model view.

2. Click on the table header or field name.

3. Add a full explanation of what the table or field contains in the description area of the Properties pane. You can see an example of this in Figure 3-2.

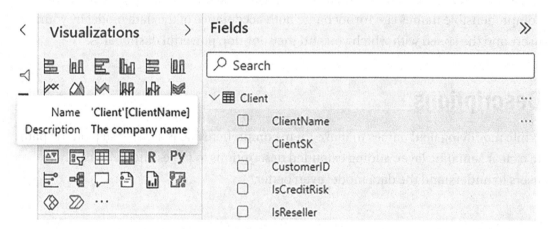

Figure 3-2. *Adding a description to a table or field*

The description that you just added will now be visible in the Report view when you hover the pointer over the table or field. You can see this in Figure 3-3.

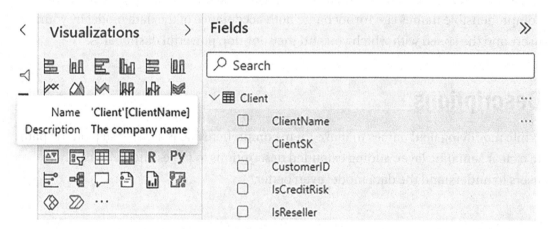

Figure 3-3. *Displaying the description to a table or field*

Hiding Tables

You do not have to remove tables to prevent them appearing in a data model. In some cases it may be more useful (and in some cases necessary) to hide the tables - which leave them and the data they contain - available inside the data model. This way certain fields can still be used in visuals even if the source data is not immediately visible.

To hide a table

1. In the Model canvas click on the visibility icon for the table that you want to hide as shown in Figure 3-4.

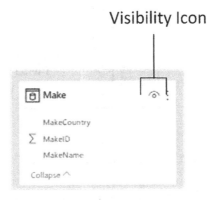

Figure 3-4. *The hide table icon*

The table will remain visible in the Data and Model views but will not appear in the Fields pane in the Report view.

You can make a table reappear by clicking the visibility icon a second time.

Tables can be hidden (and made to reappear) in other ways, too. You can

- Right-click on a table name in the Fields pane and select the Hide option. A check mark to the left of the table name will indicate that the table is currently hidden.

- Select a table and then set the Is hidden button to On in the Properties pane of the Model view. Setting this to Off causes the table to reappear in the Report pane.

One option that is useful when modifying table visibility is the View hidden option in the popup menu for tables in the Fields pane. This only appears in Report view, and it allows you to see any tables that are currently hidden.

Hiding Fields

As was the case for tables, you do not have to remove fields to prevent them appearing in a data model. In some cases it may be more useful to hide the fields - which leave them and the data they contain - available inside the data model. This way certain fields can still be used in visuals even if the source data is not immediately visible.

To hide a field

1. In the Model canvas click on the visibility icon for the field that you want to hide as shown in Figure 3-5.

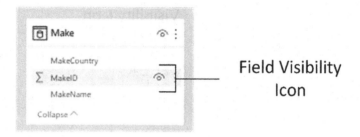

Figure 3-5. *The hide field icon*

The field will remain visible in the Data and Model views but will not appear in the Fields pane in the Report view.

You can make a field reappear by clicking the visibility icon a second time.

Fields can be hidden (and made to reappear) in other ways, too. You can

- Right-click on a field name in the Fields pane and select the Hide option. A check mark to the left of the field name will indicate that the field is currently hidden.

- Select a field inside the relevant table and then set the Is hidden button to On in the Properties pane of the Model view. Setting this to Off causes the field to reappear in the Report pane.

One option that is useful when modifying field visibility is the View hidden option in the popup menu for a table in the Fields pane. This only appears in Report view, and it allows you to see any fields that are currently hidden.

Finally, one option that is available in all Power BI views is the Unhide all option from the table context menu of the Fields pane. This will cause all hidden tables and fields to become visible.

Display Folders and Subfolders for Measures

It is not only possible, but a frequent occurrence, to have tables that contain hundreds of fields – all in alphabetical order. This can make users' lives harder than they need to be when creating dashboards.

Power BI Desktop's answer is to assemble related fields in folders. Not only can these folders contain any of the fields in a table, they can be expanded or collapsed just like folders in, say, Microsoft Outlook.

Once again we are only looking at the organization and presentation of attributes in a data model. You need to be aware that adding folders only changes the *presentation* of the model, not the model itself.

Creating a Folder

To add fields to a folder

1. Switch to Model view.

2. In the Fields pane, ctrl-click on the field(s) that you wish to add to a folder.

3. In the Properties pane add a name for the new folder that you want to create.

4. Press Enter.

In Figure 3-6 you can see an example of a folder in the Vehicle table named Model Information. Clicking the chevron to the left of the folder name will expand or collapse the folder.

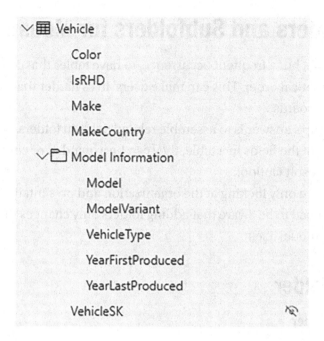

Figure 3-6. *Creating a display folder*

What is interesting to note is that you never create a blank folder in Power BI Desktop. Instead you take an existing measure (these are DAX calculations in Power BI that are explained starting at Chapter 8) and place it inside a newly defined folder. This creates the folder as well as placing the measure inside the folder.

Some things to know about folder names are that they

- Can contain numbers and/or letters

- Can begin with numbers and/or letters

- Can contain most special characters – though as a matter of good practice these are usually best avoided

Removing Fields from a Folder

To remove fields from a folder

1. Switch to Model view.

2. In the Fields pane, ctrl-click on the field(s) that you wish to remove from a folder.

3. In the Properties pane delete the folder name that is displayed.

The selected fields will reappear at the top level (the root) of the table that contains the measure (and that used to contain the display folder) in the Fields pane. This will not delete the field itself.

Moving Fields Between Folders

It is really very simple to move fields between folders.

1. Switch to Model view.

2. In the Fields pane, ctrl-click on the field(s) that you wish to move to another folder.

3. Drag the fields into the destination folder.

Alternatively, once you have selected one or more fields, enter the destination folder name the Properties pane. This can be either a totally new folder or the name of an existing folder.

Note   When moving fields between folders, you may find it easier to copy and paste folder names. This avoids creating multiple folders with not quite identical names.

Removing a Folder

To remove a folder all you have to do is to remove the fields from a folder – as described previously in "remove fields from a folder." An empty folder will automatically be removed from the data model.

SubFolders

You can "nest" measures inside subfolders (inside subfolders inside subfolders...) equally easily. All you have to do once a measure is selected is to create the top-level folder name followed by a backslash and then the subfolder name.

As an example of this, look at the subfolder name and corresponding structure that are shown in Figure 3-7.

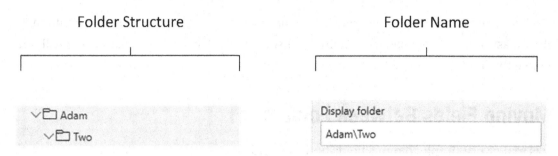

Figure 3-7. *Creating a subfolder*

Simply creating the folder path with a backslash separating the top-level and child folder names creates the two folders - or adds a subfolder to an existing parent folder. You can add multiple sub-levels simply through separating each display folder level from its parent folder by a backslash.

Items in Multiple Folders

As a final pirouette to please your users, you can also extend the semantic model so that a field appears in several folders at once.

This is achieved by (still in Model view)

1. Selecting a field or fields in the Fields Pane

2. Entering the name of the required subfolders (separated by a semicolon) in the Display folder area of the Properties pane

You can see this in Figure 3-8.

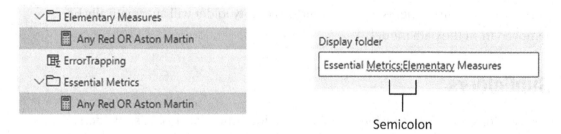

Figure 3-8. *Displaying a field in multiple subfolders*

Note   You cannot move fields between different tables – only between folders inside the table containing the field.

Tables to Contain Measures

A complex and powerful data model can contain not only hundreds of columns of data, but also hundreds of calculations (called measures). While you will be learning about measures from Chapter 8 onwards, it is nonetheless worth learning how to organize these particular elements earlier rather than later in your Power BI apprenticeship.

What Power BI can do is to let you create otherwise empty tables that exist only to host DAX measures. These tables (you are not limited to a single table) can also contain multiple display folders at multiple levels. This provides a robust and structured way to classify the DAX calculations you add to a core data model.

At its simplest, what you have to do is

1. Switch to Report view.

2. In the Home menu click the Enter data button. The Create Table dialog will appear.

3. Enter _**Measures** as the table name. You can, of course, use any name that suits you.

4. Double-click the column title and enter **Dummy** as the table name. The Create Table dialog will look like Figure 3-9.

Create Table

	Dummy	+
1		
+		

Name: _Measures

Load   Edit   Cancel

Figure 3-9. *The Create Table dialog*

    5.   Click Load.

The new, empty table will be created in the Fields pane as you can see in Figure 3-10.

Figure 3-10. *A table to contain measures*

You can now create display folders inside this table and use it to hold DAX measures that you create. This avoids adding measures to the tables that contain data.

At this point you may be wondering why you should go to the trouble of creating a table to store DAX measures. Put simply, this approach allows you to separate measures from the rest of the data model. This means that if you later change the data model (by replacing or removing tables, for instance) you are not deleting measures as well.

You can call the table anything that you want, of course. My habit is to differentiate measure tables through using a specific naming convention which is to start measure table names with an underscore. This way measure tables appear at the top of the list of tables in the Fields Pane.

Note   I realize that I am explaining folders (which contain measures) before even explaining what measures are. Measures are one type of DAX calculation. You will start to learn about these in Chapter 8.

Hierarchies

In a data model it is perfectly usual to have attributes that form natural hierarchies. In the sample data file PrestigeCarsDimensional.pbix, for instance, the vehicle Make forms the top level of a hierarchy, while model forms the next level of the hierarchy. This is because (hopefully) models are associated with one make only. Tweaking the data model to display fields in hierarchical structures like these can greatly assist users when they are analyzing the data.

It is worth underlining that we are only looking at the organization and presentation of attributes in a data model. Adding hierarchies only changes the presentation of the model, not the model itself.

Creating a Hierarchy

To create a hierarchy

1. In the Fields pane (in Report, Data, or Model view) select the field that will form the top level of the hierarchy.

2. Click on the ellipses to the right of the field name (or right-click on the field).

3. Select Create hierarchy from the context menu.

4. Click on the ellipses to the right of the hierarchy name and select Rename.

5. Enter a new name for the hierarchy and press Enter.

6. Drag the field representing the second level of the hierarchy over the hierarchy name. The second field will be added as the next level in the hierarchy.

7. Add any further levels to the hierarchy by dragging other fields over the hierarchy name.

8. Ctrl-click to select all the fields that appear in the hierarchy that remain outside the hierarchy.

9. Click on the ellipses to the right of the field name (or one of the fields) and select Hide in report view.

Note   When creating a hierarchy, it is important to add the fields in the "top down" order in which they will appear in the hierarchy.

You can see an example of a hierarchy using the sample data model PrestigeCarsDimensional.pbix – the Make and Model Hierarchy – in Figure 3-11.

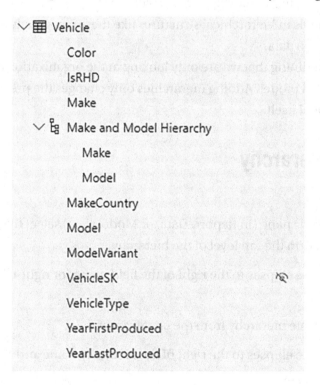

Figure 3-11. *A simple hierarchy*

A hierarchy can be expanded and collapsed just like a folder.

Removing Fields from a Hierarchy

To remove fields from a hierarchy

1. In the Fields pane (in Report, Data, or Model view), select the field that in a hierarchy you wish to remove.

2. Click on the ellipses to the right of the field name and select Delete from model. A warning dialog will appear.

3. Click OK.

Note   The warning dialog can give the impression that the field will be removed from the data model completely. This is not the case, as it is only removed from the hierarchy.

Extending a Hierarchy

You can add further levels to a hierarchy at any time.

1. In the Fields pane (in Report, Data, or Model view), find the field *in the hierarchy* where you wish to add a new sublevel.

2. Drag the field representing the new level of the hierarchy over the lowest level field hierarchy. The new field will be added as the next level in the hierarchy.

3. Click to select the original field that you added to the hierarchy that remains outside the hierarchy.

4. Click on the ellipses to the right of the field name and select Hide in report view.

Removing a Hierarchy

To remove a hierarchy that you have created

1. In the Fields pane (in Report, Data, or Model view), select the hierarchy that you wish to remove.

2. Click on the ellipses to the right of the field name and select Delete from model. A warning dialog will appear.

3. Click OK.

4. Ctrl-click to select all the fields that previously appeared in the hierarchy that remain outside the hierarchy and that you previously made invisible in Report view.

5. Click on the ellipses to the right of the field name (or one of the fields) and uncheck Hide in report view to make the fields visible again.

Note   Hierarchies can be added to folders and subfolders just as fields can.

Binning and Grouping

Sometimes you - or the users that you are developing a semantic model for - will want to avoid being drowned in detail. This can mean pre-collating data into specific groups (for data attributes) or splitting data into buckets (for values). In Power BI Desktop, these methods of data pre-preparation are called, respectively, grouping and binning.

- *Grouping* allows you to define custom groups of data attributes. This means that you can look at the values for every element in the group together.

- *Binning* places values into defined intervals reducing the number of individual values.

Groups and bins of values do not prevent you using the original values in any way. They are a supplementary way of preparing data for analysis at a less detailed level.

Grouping

Imagine that you want to break down the makes of car sold by Prestige Cars so that you can easily identify makes by their geographical origin. Grouping allows you to do this.

1. Open - or switch to - the PrestigeCarsDimensional.Pbix sample file.

2. In the Fields pane (and in the Vehicle table), right-click on the Make field (or on the ellipses for this field) to display the context menu.

3. Click New group. The Groups dialog will appear.

4. Name the group **Cars By Geographical Source**.

5. In the Ungrouped values pane on the left, ctrl-click on all the British car makes (or any makes you prefer to test the principle).

6. Click Group. This adds a group to the Groups and members pane on the right.

7. In the Groups and members pane, double-click on the name of the newly created group, and replace the list of makes with **UK Cars**.

8. In the Ungrouped values pane on the left, ctrl-click on all the European car makes (or any makes you prefer to test the principle).

9. Click Group. This adds a group to the Groups and members pane on the right.

10. In the Groups and members pane, double-click on the name of the newly created group, and replace the list of makes with **European Cars**.

11. Ensure the Include Other group check box is selected. This dialog will look like Figure 3-12.

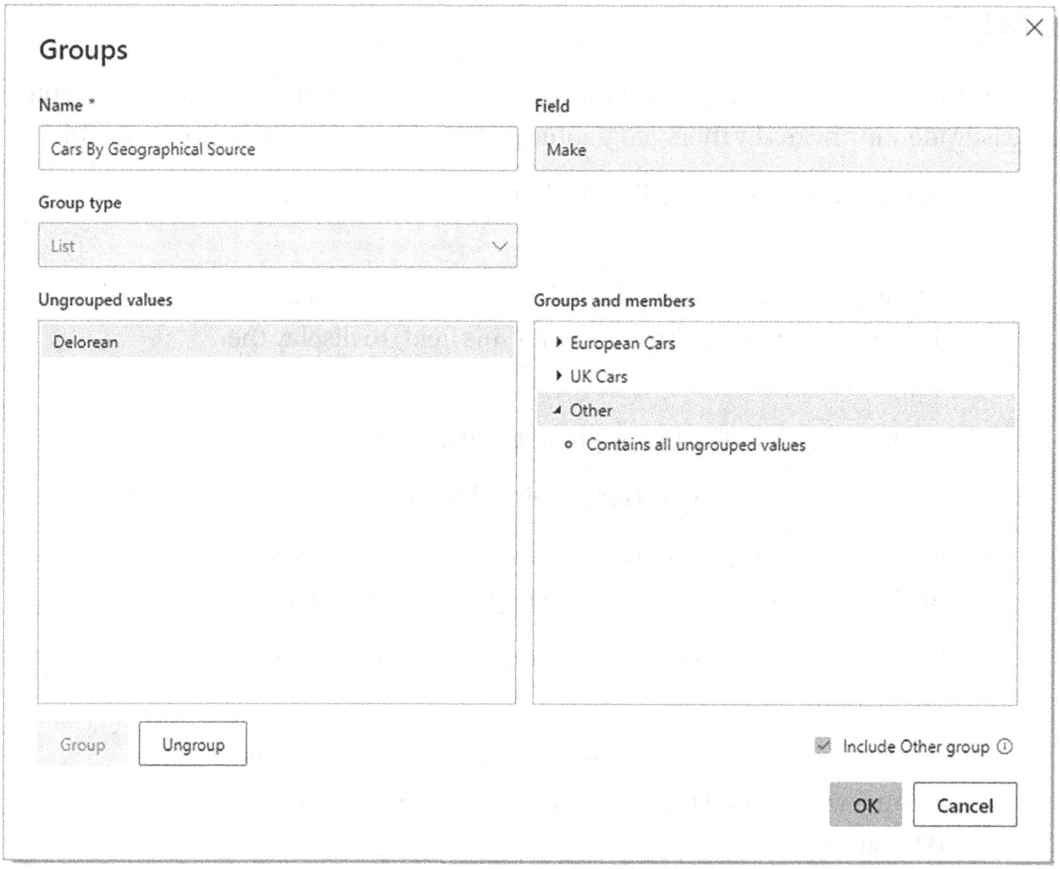

Figure 3-12. *The Groups dialog for a group of attributes*

    12.   Click OK.

The group will appear in the Fields pane inside the table where the source field is found. You can now use the group in visuals to display groups of makes of vehicle rather than individual makes. You can also use groups as filters and slicers. In the Groups dialog, you can see the fields that each group contains by clicking on the triangle to the left of each group name. This will expand and minimize the group content.

Note   Creating a group does not replace the field that you used as the basis for the grouping. Groups are an additional element in the data model. You can, of course, hide the field that underlies the group if you want only the grouped data to be available in dashboards.

Binning

Whereas grouping collates descriptive values, binning collates numeric values.
Power BI Desktop gives you two options when binning numeric data:

- Allocating data to a specified number of bins each containing approximately the same number of elements

- Creating bins at regular threshold levels

To begin with, let's group data into ten data bins.

1. Open - or switch to - the PrestigeCarsDimensional.Pbix sample file.

2. In the Fields pane (and in the Sales table), right-click on the SalePrice field (or on the ellipses for this field) to display the context menu.

3. Click New group. The Groups dialog will appear.

4. Name the group **10 Bin Sale Price**.

5. Ensure that the Group type popup is set to Bins.

6. Set the Bin type to Number of bins.

7. Set the Bin count to **10**. The dialog will look like Figure 3-13.

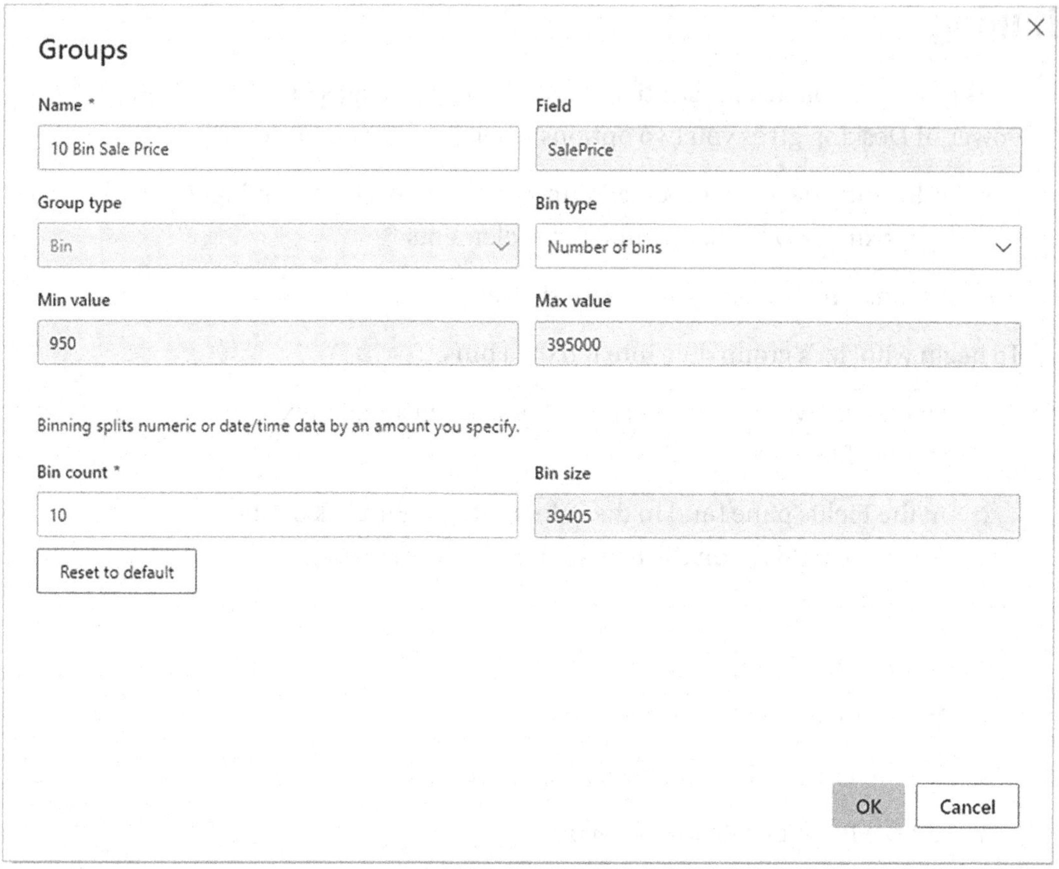

Figure 3-13. *The Groups dialog for data binning*

8. Click OK.

The 10 Bin Sale Price group will appear in the Fields pane inside the table where the source field is found. You can now use the group in visuals to display sale prices divided into ten equal bins of data rather than individual values. This can help you to get an idea of how data is distributed.

You can also, if you prefer, set bin thresholds rather than requesting a defined number of bins. Suppose that you want to see sales collated into price ranges of 0-50000, 50001-100000, etc. You can do this (still in the PrestigeCarsDimensional.Pbix sample file) by

1. In the Fields pane (and in the Sales table), right-click on the SalePrice field (or on the ellipses for this field) to display the context menu.

2.   Click New group. The Groups dialog will appear.

3.   Name the group **Sale Price Every 50000**.

4.   Ensure that the Group type popup is set to Bins.

5.   Set the Bin type to Size of bins.

6.   Set the Bin size to **50000**.

7.   Click OK.

The Sale Price Every 50000 group will appear in the Fields pane inside the table where the source field is found. You can now use the group in visuals to display sale prices divided by price range rather than individual values. This too can help you to get an idea of how data is distributed.

Note   You can modify an existing group by clicking Edit groups in the context menu for a group that you have created.

Naming Conventions in the Semantic Model

If you will be creating new metrics, then you will need to give them names. Inevitably there are a few minor limitations on the names that you can apply. So, rather than have Power BI Desktop cause problems, I prefer to explain the overall guidelines on naming conventions earlier rather than later in the course of this chapter.

The first thing to remember is that column names have to be *unique* inside each table. So you *cannot* have two columns with the *same name* inside the *same table*. You *can*, however, have two columns that share a name if they are in separate tables. However I generally advise that you try and *keep column names unique across all the tables* in a Power BI Desktop file if you possibly can. This can make building visualizations easier and safer because you do not run the risk of using a column from the "wrong" table in a chart, for instance - and getting totally inappropriate results as a consequence.

Fortunately column names are *not* case-sensitive.

Custom Formats

One of the aims of the semantic model is to ensure standardization across dashboards that use a data model. This can involve defining custom formats for values and applying them to one or more columns.

The good news (for Excel users at least) is that Power BI custom formatting is identical to Excel custom formatting. However, Power BI custom formatting is not as wide-ranging as its Excel ancestor.

To create and apply a custom format (still in the PrestigeCarsDimensional.Pbix sample file)

1. Switch to Model view, and ensure that the Properties pane is visible.

2. In the Sales table select the SalePrice field (you can select this either in the Fields Pane or the table in the data model).

3. Select Custom as the Format.

4. Enter the following custom format in the Custom format area: **"£ "#,0;-"£ "#,0;"£ "#,0**. The Properties pane will look like Figure 3-14.

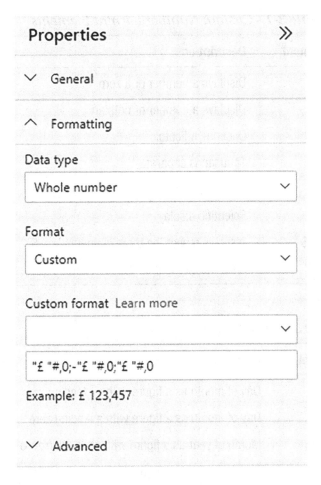

Figure 3-14. *Defining a custom format*

5. Apply this Format to any other appropriate field by copying the format string and applying the steps described above.

Tables 3-1 to 3-3 give a short reminder of Power BI custom formatting elements that you can use to define your own custom formats.

Table 3-1. *Custom Number Format Elements*

Element	Description
0	Displays a number or a zero
#	Displays a number or nothing
.	Decimal indicator
,	Thousands separator
%	Percentage sign
E	Scientific display
- + $	Displays one of the three elements: - + $
"xx"	Adds the element enclosed in double quotes

Table 3-2. *Custom Date Format Elements*

Element	Description
d	Day of month as a figure with no leading zero
dd	Day of month as a figure with a leading zero
m	Month of year as a figure with no leading zero
mm	Month of year as a figure with a leading zero
mmm	Month of year abbreviated
mmmm	Month of year full
yy	Year as two figures
yyyy	Year as four figures

Table 3-3. *Custom Time Format Elements*

Element	Description
h	Hours with no leading zero
hh	Hours with a leading zero
n	Minutes with no leading zero only if preceded by h or hh
nn	Minutes with a leading zero only if preceded by h or hh
m	Minutes with no leading zero
mm	Minutes with a leading zero
s	Seconds with no leading zero
ss	Seconds with a leading zero

A format string is in three parts, each separated by a semicolon:

- Firstly the format for positive numbers

- Secondly the format for negative numbers

- Finally the format for empty values

Data Model Aesthetics

When Power BI Desktop loads tables from data sources, it may attempt to create a readable view of the data model, but more often than not you will have to rearrange the layout of the tables to improve the clarity of the overall data picture. Now, although repositioning tables can be considered pure aesthetics, I find that adjusting the presentation of the data model is really useful. A well laid out data set design will help you understand the relationships between the tables and the inherent structure of the data.

A clean-looking data model is nearly always a sign that the model has been well thought out. Indeed, I find that trying to make a data model as clear as possible both helps when defining the model and when modifying it at a later data. So I encourage you to try and make your data models look good as this could help you significantly.

So what makes a data model look good? Essentially (and it at all possible) it means placing tables in the design canvas so that joins are clearly visible and tables that are joined are in reasonable proximity. You might not need to see all the fields that are present in a table (after all, the data model is focused on tables, not fields), but you should not let tables overlap as this obscures the view of the whole model.

To make your data model more pleasant – and instructive – to look at, you can use Model view to move and resize the tables. Moving a table is as easy as dragging the table's title bar. Resizing a table means placing the pointer over a table edge or corner and dragging the mouse. Any relationships that have been created will persist as tables are moved around the canvas in Model view.

Collapsing and Expanding Tables

One technique that you may find necessary to implement when handling larger data models is collapsing tables. Essentially this means (in Model view only) hiding some or all of the fields and only leaving the table name visible.

To collapse a table

1. In the Model view click Collapse at the bottom of the table you wish to collapse.

To expand a table

1. In the Model view click Expand at the bottom of the collapsed table you wish to expand.

You can see an example of the same table, both expanded and collapsed in Figure 3-15.

Figure 3-15. *Comparing collapsed and expanded tables*

Note   You can only resize a table that is expanded (unless you're making it wider).

While collapsing tables frees up a considerable amount of on-screen real estate, it can make understanding of the joins between tables more complex. So it is advisable, when collapsing tables, to ensure that the fields used in table relationships remain visible even once a table is collapsed.

To do this

1. In the Model view click anywhere inside the Model canvas – but not on a table, field, or relationship.

2. In the Properties pane ensure that Show related fields when a card is collapsed is set to On.

This will ensure that for all tables you can see not only the table name, but also all fields that participate in joins.

Tabs in Data Modeling

If your data model is not too large (which usually means that there are a dozen or so tables or less), then you can usually see the entire model in a single screen in Model view.

However, if your model starts becoming any larger, then it might be difficult to see the entire model legibly on a single screen – even if the tables are collapsed.

Fortunately, Power BI Desktop has a solution to this challenge. You can organize sections of your data model in separate tabs (much as your dashboards probably contain multiple tabs).

This is as easy as

1. In Model view Click the "plus" sign at the bottom of the Power BI Desktop window to the right of any existing tabs. A new tab will be created.

2. Drag the tables from the Fields pane that you wish to focus on in this new tab. They will appear in the tab and all joins will be preserved.

3. Right-click on the new tab name, and select Rename from the context menu.

4. Enter the new name.

5. Press Enter.

Any structural changes that you make to any table or field in any of the layout tabs will be reflected across all layout tabs. It is worth noting, however, that any aesthetic changes that you make will only apply to the selected tab.

You can see an example of a tab that displays only selected tables in Figure 3-16.

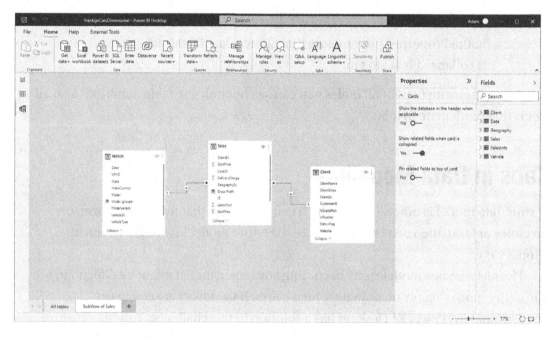

Figure 3-16. *Adding new tabs to the data model*

Note   Unlike dashboard tabs, layout tabs cannot be moved left or right.

To remove a layout tab

1. Click on the cross icon at the top right of the tab name.

2. Click Delete in the warning dialog that appears.

Removing a layout tab will not remove any tables that it contains from the data model or the All tables tab.

Note   You cannot delete the initial "All tables" tab.

Conclusion

Data models are more than just adding tables in an appropriate topology. They are also about preparing the data so that it is intuitively usable. This is where extending the structural aspects of the data model so that it becomes a semantic model is important.

So you learned how to rename fields and tables and, if necessary, hide fields that are not intended to be used. You also saw how to add descriptions to tables and fields to assist both you and users to understand the semantic model.

Then you moved on to adding elements such as hierarchies and groups. The latter can contain either attributes or bins of values.

Finally you learned how to add display folders inside tables, create custom formats, and isolate complex data models into tabs (or views) that allow you to focus on a subset of tables. You also saw how to add tables designed to hold the measures that you will be learning to develop as you move on to look at DAX - which begins in the next chapter.

CHAPTER 4

Calculated Columns

This chapter further develops the dimensional data model that you saw in the previous chapter. It will explain how to augment the existing tables that you have imported by adding new columns containing calculations. You can then apply the output from these calculated columns to the dashboards that you create using Power Bi Desktop.

Admittedly, not every data model in Power BI Desktop will need extensive additional calculations. In some cases the data can speak for itself without much polishing. Yet Business Intelligence is, at its heart, based on figures. Consequently you will need sooner or later to apply simple math, calculate percentages, or compare figures over time. You may even want to develop more complex formulas that enable you to extend your analyses and illustrate your insights. Fortunately Power BI Desktop makes these - and many, many other calculations - very easy. What is more, if you are an Excel user, you will probably find many if not most of the techniques that are explained in the four chapters devoted to adding calculated columns to the data model totally intuitive.

In some cases you will only need to cherry-pick techniques from the range of available options to finalize a dataset. So it probably helps to know what Power BI Desktop can do, and when to use the techniques that are available. Therefore I leave it to you to decide what is fundamental and what is useful. The objective in the chapters devoted to adding calculated columns is to present a tried and tested suite of core calculation solutions so you are empowered to deal with the range of the potential challenges that you may encounter in your data analysis.

All calculations in the Power BI Desktop data model are written using a language named DAX. This stands for **D**ata **A**nalysis e**X**pressions. As you will see, DAX is not in any way a complex programming language and is indeed known as a *formula language* because all it is a set of some 300+ formulas that you can use and combine to extend data models and create the metrics that you need to underpin the visualizations in the dashboards that you create. Fortunately DAX formulas are loosely based on the formulas in Excel (indeed nearly a third of DAX functions are identical to Excel functions are identical), so the learning curve for an Excel power user can be quite short.

© Adam Aspin 2023
A. Aspin, *Pro DAX and Data Modeling in Power BI*, https://doi.org/10.1007/978-1-4842-8995-2_4

Given the vast horizons that DAX opens up to Power BI Desktop users, the remainder of this book will be required to explain many of the more frequently used DAX functions. To apply some structure to a potentially huge and amorphous area, the introduction to DAX is divided into six parts:

- This chapter along with Chapters 5 through 7 covers *column-based calculations* (where the formula appears as a new column in a table).

- Chapters 8 through 12 start an introduction to *measures* (calculations that are added to a table but that do not add a column of calculated data).

- Chapters 13 and 14 describe how to create a Date dimensions and add *time intelligence* (measures that are used to aggregate data over time periods or to compare data over time).

- Chapter 15 introduces DAX variables.

- Chapter 16 introduces DAX Table functions.

- Chapter 17 explains how to push the data model to new levels through using DAX to go beyond the initial limitations of relationships and cross-filters.

- Chapter 18 explains the all-important concept of context in calculations.

If you want to continue enhancing the data as it appeared at the end of the previous chapter, download the PrestigeCarsDimensional.pbix file from the Apress website. This file will let you follow the examples as they appear in this chapter.

Types of Calculations

If you are really lucky, then the data that you have imported contains everything that you need to create all the visualizations you can dream up in Power BI Desktop. Reality, however, is frequently more brutal than that, and it necessitates adding further metrics to one or more tables. These calculated metrics will extend the data available for visualization. This is fundamental when you are using tools such as Power BI Desktop that do not allow you to add calculated elements to the visuals directly, but insist that all metrics — whether they are source data or calculated metrics — exist in the dataset. This is less of a constraint and more of a nod toward good design practice, because it forces you to develop calculations once and to place them in a single central repository. It also reduces the risk of error, because this way users can be guided not to develop their own (possibly erroneous) metrics and calculations and so distort the truth behind the data. Instead they can use metrics that you have designed for them to use.

When creating DAX metrics, you will be defining elements that are of practical use when creating dashboard visualizations. This can include

- Creating derived metrics that will appear in visualizations

- Adding elements that you will use to filter pages or visualizations

- Creating elements that you use to segment or classify data

- Defining new metrics based on existing metrics

- Adding your own specific calculations (such as accounting or financial formulas)

- Adding weightings to values

- Ranking and ordering data

- And many, many more...

Calculated Columns or Measures?

From the start it is essential to understand that there are two types of DAX calculations that you can write:

- Calculated columns

- Measures

The two approaches are somewhat different both in how they are written and how they are used. As calculated columns are generally simpler and easier, I prefer to explain these first as they offer a gentler learning curve. A little later in this book (starting at Chapter 8), you will start to learn measures and then progress to see how and when you should use one or the other technique when extending the data model with DAX calculations.

Calculated Columns

Adding calculated columns is a key way in which you can extend a dataset with derived metrics that you can use in Power BI Desktop dashboards. There are multiple reasons why you may need to develop the data model by adding further columns. Some reasons, among many, many others, are

- Concatenating data from two existing columns into one new column

- Performing basic calculations for every row in the table, such as adding or subtracting the data in two or more columns

- Extracting date elements such as the month or year from a date column and adding them as a new column

- Extracting part of the data in a column into another column

- Replacing part of the data in a column with data from another column

- Creating the column needed to apply a visually coherent sort order to an existing column

- Showing a value from a column in a linked table inside the source table

Indeed, the list could go on.

Before you start wondering exactly what you are getting yourself into, I want to add a few words of reassurance about the ways in which a data model can be extended.

- **First,** extending a table with added columns is designed to be extremely similar to what you would do in Excel. Consequently you are in all probability building on your existing knowledge as a spreadsheet power user.

- **Second**, the functions that you will be using are, wherever possible, similar to existing Excel functions. This does not mean that you have to be an Excel Super User to add a column but that knowledge gained using Excel will help with Power BI Desktop and vice-versa.

- **Finally**, most of the basic table extension techniques follow similar patterns and are not complex in themselves. So the more you work at adding columns, the easier it will become as you re-use and extend techniques and formulas.

Creating columns is a bit like creating a formula in Excel that you then copy down over the entire column for a list or table. They are even closer to the derived columns that you can add to queries in Microsoft Access. The key thing to note is that any formula will be applied to the *entire* column.

It is worth noting from the start that a formula that you add to a new column will be calculated and applied to a column when it is created. It will only be *recalculated* if you refresh the entire table or Power BI Desktop file.

Note   You do not need to be an experienced Excel user to write DAX. However I want to reassure Excel users that they can capitalize on their acquired knowledge and skills when moving from Excel formulas to DAX.

How to Add a Calculated Column

There are a few ways that you can add a calculated column in Power BI Desktop. However, when starting out on the road to mastering DAX I suggest the following approach:

1. Switch to Data View using the Data icon on the top left of the Power BI Desktop Window.

2. In the Fields pane, click on the table name where you want to add the column.

3. Click on the New column button in the Table tools menu (normally this menu is activated by default).

For reference, other ways that you can add a calculated column are

- In either Report or Data view, right-click on the table in the Fields pane where you want to add the calculated column, and select New column from the context menu.

- In either Report or Data view, click on the ellipses to the right of the table name in the Fields pane where you want to add the calculated column, and select New column from the context menu.

Concatenating Column Contents

As an initial example of a Calculated Column, I will presume that, when working with Power BI Desktop, you have met a need for a single column of data that contains both the make and model of every car sold. Because the data we imported contains this information as separate columns, we need to add a new column that takes the data from the columns Make and Model, and joins them together (or concatenates them, if you prefer) in a new column. Here is how it can be done:

1. Open the file PrestigeCarsDimensional.pbix from the folder where you copied the source data from the Apress website.

2. In the Power BI Desktop window, make sure that you are in Data View.

3. Click on the Vehicle table in the Fields pane. The first rows of this table will be displayed.

4. In the Table tools ribbon, click the New Column button. A new empty column will appear to the right of the final column of data. This column is currently entitled *Column* and is highlighted as shown in Figure 4-1.

VehicleSK	Make	Model	Color	ModelVariant	MakeCountry	YearFirstProduced	YearLastProduced	VehicleType	IsRHD	Column
14	Aston Martin	DB2	Black	NULL	GBR	NULL	NULL	Coupe	1	
15	Aston Martin	DB2	Black	NULL	GBR	NULL	NULL	Saloon	1	
16	Aston Martin	DB2	Blue	NULL	GBR	NULL	NULL	Saloon	1	
17	Aston Martin	DB2	British Racing Green	NULL	GBR	NULL	NULL	Saloon	1	
18	Aston Martin	DB2	Canary Yellow	NULL	GBR	NULL	NULL	Saloon	0	
19	Aston Martin	DB2	Green	NULL	GBR	NULL	NULL	Saloon	0	
20	Aston Martin	DB2	Green	NULL	GBR	NULL	NULL	Saloon	1	
21	Aston Martin	DB2	Night Blue	NULL	GBR	NULL	NULL	Saloon	1	
22	Aston Martin	DB2	Red	NULL	GBR	NULL	NULL	Saloon	1	
23	Aston Martin	DB2	Silver	NULL	GBR	NULL	NULL	Saloon	1	
24	Aston Martin	DB4	Black	NULL	GBR	NULL	NULL	Saloon	1	
25	Aston Martin	DB4	Canary Yellow	NULL	GBR	NULL	NULL	Saloon	1	
26	Aston Martin	DB4	Green	NULL	GBR	NULL	NULL	Saloon	1	

Figure 4-1. *A new, blank calculated column*

5. The formula bar above the table of data will display `Column = .`

6. In the Formula bar click to the right of the equals sign. Press [. A list of the fields in the current table will appear. This will look like Figure 4-2.

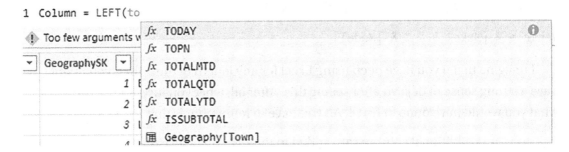

Figure 4-2. *Selecting a field for a formula*

7. Click on the field [Make].The formula bar now reads =[Make].

8. In the formula bar add: **& " " &.** The formula bar now reads
   Column =[Make] & " " &.

9. Press [. And select [Model] from the list of fields. The formula bar
   now reads Column = [Make] & " " & [Model].

10. Press Enter (or click the tick icon in the Formula bar). The column
    is filled automatically with the result of the formula and shows the
    make and model of each car sold.

11. Right-click on the column header for the new column and
    select Rename.

12. Enter the word **Vehicle** and press Enter.

The table will now look something like Figure 4-3. Moreover the new column has
been added as a field to the Vehicle table in the Fields list.

VehicleSK	Make	Model	Color	ModelVariant	MakeCountry	YearFirstProduced	YearLastProduced	VehicleType	IsRHD	Vehicle
14	Aston Martin	DB2	Black	NULL	GBR	NULL	NULL	Coupe	1	Aston Martin DB2
15	Aston Martin	DB2	Black	NULL	GBR	NULL	NULL	Saloon	1	Aston Martin DB2
16	Aston Martin	DB2	Blue	NULL	GBR	NULL	NULL	Saloon	1	Aston Martin DB2
17	Aston Martin	DB2	British Racing Green	NULL	GBR	NULL	NULL	Saloon	1	Aston Martin DB2
18	Aston Martin	DB2	Canary Yellow	NULL	GBR	NULL	NULL	Saloon	0	Aston Martin DB2
19	Aston Martin	DB2	Green	NULL	GBR	NULL	NULL	Saloon	0	Aston Martin DB2
20	Aston Martin	DB2	Green	NULL	GBR	NULL	NULL	Saloon	1	Aston Martin DB2
21	Aston Martin	DB2	Night Blue	NULL	GBR	NULL	NULL	Saloon	1	Aston Martin DB2
22	Aston Martin	DB2	Red	NULL	GBR	NULL	NULL	Saloon	1	Aston Martin DB2
23	Aston Martin	DB2	Silver	NULL	GBR	NULL	NULL	Saloon	1	Aston Martin DB2
24	Aston Martin	DB4	Black	NULL	GBR	NULL	NULL	Saloon	1	Aston Martin DB4
25	Aston Martin	DB4	Canary Yellow	NULL	GBR	NULL	NULL	Saloon	1	Aston Martin DB4
26	Aston Martin	DB4	Green	NULL	GBR	NULL	NULL	Saloon	1	Aston Martin DB4
27	Aston Martin	DB4	Night Blue	NULL	GBR	NULL	NULL	Coupe	1	Aston Martin DB4
28	Aston Martin	DB4	Pink	NULL	GBR	NULL	NULL	Saloon	1	Aston Martin DB4

Figure 4-3. *An initial calculated column*

The formula that you entered will finally read

```
Vehicle = [Make] & " " & [Model]
```

I imagine that if you have been using Excel for any length of time, then you might have a strong sense of déjà vu after seeing this. After all, what you just did is virtually what you would have done in Excel. All you have to remember is that

- Any additional columns are added to the *right* of the existing columns. You *cannot* move them elsewhere in the table once they have been created.

- Column names will always appear in alphabetical order in the Fields pane. All functions begin with the equals sign.

- Any function can be developed and edited in the formula bar at the top of the table.

- Reference is always made to *columns*, **not** *to cells* (as you would in Excel).

- Column names are always enclosed in square brackets.

- You can nest calculations in parentheses to force inner calculations before outer calculations — again, just as you would in Excel.

Once a new column has been created, it will remain at the right of any imported columns in the table where you added it. The field that it represents will always be added in alphabetical order to the collection of fields for this table in the Fields List.

If you look closely at the field that was added to the Fields pane, you will notice that there is a tiny icon of a table with Fx to its left. You can see this in Figure 4-4.

Figure 4-4. *The icon for a calculated column*

This is how you can distinguish new (or "calculated") columns from other fields such as numeric fields (which have a sigma-Σ-icon) or measures (that you will meet in Chapter 8 and that have a small calculator icon) to their left.

Note   In this example you selected columns from the popup list of the available columns in the table. You can enter the column name in the Formula Bar if you prefer, but if you do then you *must* enclose the column name in square brackets. You must also enter it *exactly* as it appears in the Fields List and column title.

Renaming Calculated Columns

In the previous section I explained one of the ways that you can rename a calculated column. In fact, when it comes to renaming columns (or indeed measures), you are spoilt for choice. The currently available methods are

- Right-click on the field (column) name in the Fields pane, and select Rename.

- Click on the ellipses to the right of the field (column) name in the Fields pane, and select Rename.

- When creating the formula, replace the word Column in the formula bar with the name to use.

- Click on the field (column) name in the Fields pane, and modify the column name (to the left of the initial equals sign) in the formula bar.

- Double click the column header.

- Double click the field name.

- Press F2 when the column is selected.

- Press F2 when the field is selected.

- Use the "column tools" ribbon when the column is selected.

Tweaking Text

DAX will also let you clean up and modify the text in the tables that you have imported into your data model. Indeed, it offers a wide range of functions that you can apply to standardize and cleanse text in tables. As an example let's imagine that you want to create a column in the Clients table that contains a shortened version of each town. In fact what you want is to extract a three-letter acronym from the first three letters of the town name that you can use later in charts.

1. In the Power BI Desktop window, make sure that you are in data view.

2. In the Fields pane, click on the Geography table. This will display the data for the Geography table.

3. In the Table tools ribbon, click the New Column button. A new empty column will appear to the right of the final column of data. This column is currently entitled Column and is highlighted.

4. The formula bar above the table of data will display Column = .

5. In the Formula bar click to the right of the equals sign.

6. Type **LEFT(** Once the function appears in the popup menu, you can select it or press the Tab key if you prefer instead of typing it in its entirety. As you can see in Figure 4-5, the formula will appear in the popup list as you type.

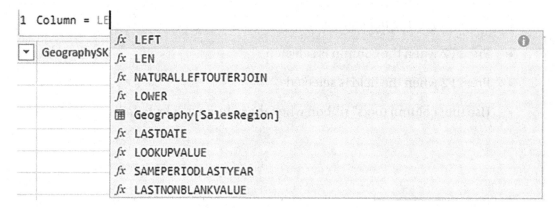

Figure 4-5. *The popup list of DAX functions*

7. Start typing the field name that you wish to insert (town in this example). After entering a few characters, you will see the field name in the popup list as shown in Figure 4-6.

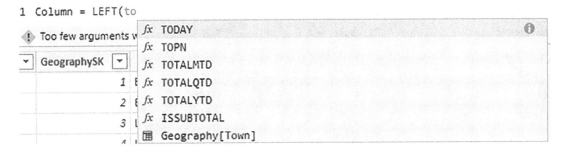

Figure 4-6. *Selecting a field from the popup list in the Formula bar*

8. Select the appropriate field (Town). Geography[Town] will appear in the Formula Bar.

9. Enter a comma.

10. Enter the figure **3**. This indicates to the LEFT( ) function that it is the *three* characters on the left that you want to isolate for each row in this column.

11. Add a right parenthesis.

12. Still inside the Formula Bar, replace Column with **TownAbbreviation**. The formula Bar will read

```
TownAbbrevation = LEFT(Geography[Town],3)
```

Press Enter (or click the tick icon in the Formula bar). The column is filled automatically with the result of the formula and shows the first three characters of every town. This is shown in Figure 4-7.

	1 TownAbbrevation = LEFT(Geography[Town],3)

CountryName ▼	GeographySK ▼	Town ▼	PostCode ▼	SalesRegion ▼	TownAbbrevation ▼
Germany	1	Berlin		EMEA	Ber
Belgium	2	Brussels		EMEA	Bru
Switzerland	3	Lausanne		EMEA	Lau
United States	4	Los Angeles		North America	Los
France	5	Lyon		EMEA	Lyo
Spain	6	Madrid		EMEA	Mad
France	7	Marseille		EMEA	Mar
Italy	8	Milan		EMEA	Mil
United States	9	New York		North America	New
France	10	Paris		EMEA	Par
Germany	11	Stuttgart		EMEA	Stu
United States	12	Washington		North America	Was
Italy	13	Rome	120	EMEA	Rom
Italy	14	Rome	129	EMEA	Rom

Figure 4-7. *A calculated column using the LEFT( ) DAX function*

As you can see, the LEFT( ) function takes two parameters:

- *Firstly* - the field from which you want to extract the leftmost characters.

- *Secondly* - the number of characters to extract.

And that is all that you have to do. By applying a simple text formula, you have prepared a column of text for effective use in a visualization.

Note   You could have selected the field to use inside the LEFT() function by entering [and selecting the field name as you did in the previous section.

DAX contains a couple of dozen functions that you can apply to the text in columns. Most of them follow the same principle as the LEFT( ) function in that they take at least two parameters, the first of which is the column that you want to take as the basis for your new column and the second (or even third) parameters provide information about how the modification is to be applied. As I do not have space to explain each and every one of these functions, Table 4-1 contains a succinct overview of a selection of some of the most useful text functions. This table does not explain all the subtleties of every function, but is destined to be both a brief introduction and a starting point for your DAX formulas that rework the text elements of your data tables.

Table 4-1. *Core Power BI Desktop Text Functions*

Function	Description	Example
LEFT()	Extracts a specified number of characters from the left of a column	LEFT(Geography[Town], 3)
RIGHT()	Extracts a specified number of characters from the right of a column	RIGHT(Vehicle[MakeCountry], 2)
MID()	Extracts a specified number of characters (the second parameter) from a specified position defined by the number of characters from the left (the first parameter) inside a column	MID(Vehicle[MakeCountry], 1, 1)
UPPER()	Converts the data to uppercase	UPPER(Client[ClientName])
LOWER()	Converts the data to lowercase. This function takes no parameters	LOWER(Client[ClientName])
TRIM()	Removes any extra spaces (trailing or leading) from the text inside a column. This function takes no parameters	TRIM(Client[ClientName])
LEN()	Counts the number of characters in a column. This is often used with the MID() function. This function takes no parameters	LEN(Client[ClientName])

(continued)

Table 4-1. (*continued*)

Function	Description	Example
FIND()	Gives the starting point (as a number of characters) of a string inside a column. This function is case-sensitive. Interestingly the first parameter is the text to find and the second is the column	`FIND('Car',Client[ClientName])`
SEARCH()	Gives the starting point (as a number of characters) of a string inside a column. This function is not case-sensitive. Interestingly the first parameter is the text to find and the second is the column	`SEARCH('Car',Client[ClientName])`
SUBSTITUTE()	Replaces one text with another inside the column. This is a bit like the search and replace function in a word processor	`SUBSTITUTE(Client[ClientName], 'Car', 'Vehicle')`
VALUE()	Converts a figure in a text column to a numeric data type. This function takes no parameters	`VALUE(Client[CustomerID])`
FIXED()	Takes a number and rounds it to a specified number of decimals then converts it to a text. The second parameter indicates the number of decimals to apply	`FIXED(Dales[LaborCost], 2)`

There are a couple of points to note now that you have seen how to use DAX formulas in Power BI Desktop:

- You can paste DAX formulas into the Formula bar if you are "lifting" DAX from sources such as this book.

- You can enter functions in uppercase or lowercase (or even a mixture of both if you want).

I imagine by now that you are feeling that DAX formulas are not only relatively easy, but also made easier by their close relationship to Excel formulas. So let's move on to see a few more.

Using Table Names in Calculated Columns

As you saw in the previous sections, when you select a field name (or if you enter it), Power BI Desktop can display the field name on one of two ways:

- The field name on its own (but enclosed in square brackets)

- The field name enclosed in square brackets and preceded by the table name

You can decide whether you wish to add the table name before the field name. In practice you have three ways of adding a field name:

- Enter the field name manually (with or without a table name).

- Start by entering a [- then select the field name from the list of fields in the popup (this will not add the table name).

- Type a few characters from the field name then select the field name from the list of fields in the popup (this will add the table name).

It is generally considered good practice to use *both the table and field names* in DAX formulas as this can make understanding - and debugging – formulas easier later.

Note   If the table name contains spaces or allowable special characters, then the table name must be enclosed in *single* quotes. There should never be a space (or anything else between the table and field names).

Handling Mistakes

It is all too easy to make mistakes when learning how to add calculated columns. After all, this is part of the learning curve. The trick is to know how to deal with any errors that you may inadvertently make.

If the formula has not been developed – or confirmed – then you can back out by pressing Escape or clicking the cross icon in the Formula bar.

If you have confirmed a formula that contains an error, you could see the new column added to the table – but with a warning icon, as shown in Figure 4-8.

Figure 4-8. *A calculated column containing an error*

Here the easy solution is to delete the column which has been added to the active table just as you would for any field. You can then try and rebuild the calculated column from scratch. Of course, if you have a more complex formula and can understand where the problem lies, you can edit the formula to correct the issue.

Modifying a Calculated Column

If you are modifying a calculated column, then it can be as simple as

1. In the Power BI Desktop window, make sure that you are in data view.

2. In the Fields pane click on the column to modify. The formula bar will appear containing the current DAX formula.

3. Edit the formula.

4. Click the tick icon in the formula bar.

Simple Calculations

To extend the basic principle, and also to show a couple of variations on a theme, let's now add a calculation to the Sales table. More precisely, I will assume that our Power BI Desktop visualizations will frequently need to display the figure for the direct costs relating to all vehicles purchased, which I will define as being the purchase price plus any related costs. Here is how you can enter a calculation like this:

1. In Data View select the Sales table.

2. Click New Column in the Table tools ribbon.

3. Replace the word Column in the formula bar with the word Direct Costs (notice that there is a space, because column names can contain spaces).

4. To the right of the equals sign, enter a left square bracket [. The list of the fields available in the Vehicle table will appear in the formula bar.

5. Type the first few characters of the column that you want to reference — CostPrice in this example. The more characters you type, the fewer columns will be displayed in the list.

6. Click on the column name. It will appear in the formula bar (including the right bracket).

7. Enter the plus sign.

8. Enter a left square bracket [and select the column SpareParts.

9. Enter a plus.

10. Enter a left square bracket [and select the column LaborCost. The formula should read

    ```
    Direct Costs = [CostPrice] + [SpareParts] + [LaborCost]
    ```

11. Click the tick box in the formula bar (or press Enter) and the new column will be created. It will also appear as a new field in the Fields list for this table.

As you can see, using arithmetic in calculated columns in Power BI Desktop is almost the same as using calculating cells in Excel. If anything it is easier as you do not have to copy the formula down over hundreds or even thousands of rows as the formula will automatically be applied to every row in the table. You can see this in Figure 4-9 (where the formula has also been formatted using the techniques you learned in Chapter 2).

DirectCosts
£8,886
£5,234
£3,849
£11,450
£8,588
£5,738
£29,810
£34,804
£12,303
£12,206

Figure 4-9. *A simple calculated column*

Note   You can include spaces in formulas to add clarity if you want.

If you are a PowerPivot user, then the sense of déjà-vu is probably so total as to be overwhelming. In fact most PowerPivot users will probably only need to skim through this chapter as the Data View of Power BI Desktop is very similar to the PowerPivot window in Excel. Except for the tables that now appear on the right instead of tabs at the bottom of the window and the current absence of a formula button, there are few differences except for the overall aesthetics. So feel free to jump over any sections that you know by heart already in this chapter (or the next) if you are a PowerPivot expert.

Tip   You can include spaces in column names if you want. After all, this is how the name will appear in your dashboards. Ideally, of course, all names should be clear for business users and add value to the semantic model.

Math Operators

For the sake of completeness, and in case there are any newcomers to the world of Microsoft products out there, I prefer to recapitulate the core math functions that are available in Power BI Desktop. These are given in Table 4-2.

Table 4-2. *Core Power BI Desktop Math Operators*

Operator	Description	Example
+	Adds two elements	[SpareParts] + [LaborCost]
-	Subtracts one element from another	[CostPrice] - [SpareParts]
/	Divides one element by another	[CostPrice] / [SpareParts]
*	Multiplies one element by another	[CostPrice] * 1.5
^	Raises one element to the power of another	[CostPrice] ^ 2

If you are working in Business Intelligence, then you are certainly able to perform basic math operations. Consequently I will not re-explain things you most likely already know. Just use the same arithmetical operators as you would use in Excel and, after a little practice, you should be able to produce calculated columns with ease. Remember nonetheless that you have to enclose in parentheses any part of a formula that you want to have calculated before the remainder of the formula. This way you will avoid any unexpected results in your dashboards.

More Complex Math

I expect that virtually all Power BI Desktop users are already proficient in Excel. So it will come as no surprise that more complex math in DAX follows the core principles that you are (probably) already used to in your spreadsheet analytics. In other words, the key to flawless DAX math is to use parentheses around any parts of the formula that you wish to calculate first.

As a simple example, suppose that you want to calculate (in a single column) the projection of what a 9% increase in direct costs will do to the gross margin. In other words you want to

1. Add up the cost price, spare parts cost, and labor cost

2. Multiply this by 1.09

3. Subtract this total from the sale price

Of course, this is not difficult, and it can be done with the following calculated column added to the Sales table

```
Reduced Margin =
[SalePrice] - (([CostPrice] + [SpareParts] + [LaborCost]) * 1.09)
```

The art in formulas like these is to wrap the lowest levels of calculation in parentheses to force each level to be calculated before another level of calculation is applied. Consequently, this example

1. Wraps the cost price, spare parts cost, and labor cost calculation inside parentheses

2. Encases the percentage calculation of the previous calculation inside a second set of parentheses

3. Subtract this total from the sale price (as this is the outer level no parentheses are needed)

You need to remember not to over-complicate formulas by adding parentheses when they are not strictly necessary as this only makes the DAX harder to read. If all you are doing is adding and subtracting, then no parentheses are needed. The following calculated column that shows the margin on each sale can be defined without the added clutter of pointless parentheses.

```
Margin = [SalePrice] - [CostPrice] - [SpareParts] - [LaborCost]
```

You can see the columns for Margin and Reduced Margin in Figure 4-10.

Margin ▼
5114
3266
2301
3550
3862
2262
7690
13196
4787
6794
6554
20114
2998

Reduced Margin ▼
1814.26
294.94
-545.41
19.4999999999982
589.08
-754.42
2507.1
7563.64
1179.73
3195.46
2663.86
13304.26
-312.18

Figure 4-10. *More complex calculations in calculated columns*

Note   Although you just saw how to create percentages in a calculated column, this is probably not something that you would do in practice very often. This is because (among other things) when the contents of the column are used in visuals, they will be aggregated – and the total of a column of percentages is rarely very useful. Chapter 10 will explain how to create robust and reusable percentage totals in detail using DAX measures.

Rounding Values

You can round and truncate values when you are preparing data ready for loading into a data model. In practice, of course, you might not yet be aware that you need to tweak your data at this stage. Fortunately DAX also contains a range of functions that can be used to round values up and down, or even to the nearest hundred, thousand, or million if need be.

As an example of this I will take the column Direct Costs that you created previously and will round it to the nearest integer. This way you will also learn how to modify a formula in DAX.

1. In Data View select the Sales table.

2. Click inside the Direct Costs column. The column will be selected, and the formula that you created previously will appear in the Formula Bar. You can see this in Figure 4-11.

Figure 4-11. *The formula bar used to edit an existing DAX calculation*

3. Click inside the Formula Bar to the right of the equals sign.

4. Enter **ROUND(**. You can also select the formula from the popup if you prefer.

5. Click at the right of the formula in the Formula Bar.

6. Enter a comma.

7. Enter a **0**.

Add a right parenthesis to complete the Round() function. You will see that the corresponding left parenthesis is highlighted in the Formula Bar to help you track which pair of parentheses is which. The formula should read

```
Direct Costs =
ROUND([CostPrice] - ([SpareParts] + [LaborCost]), 0)
```

8. Click the tick box in the formula bar (or press Enter). The formula will be modified and any decimals removed from the data in the column.

This example introduced the ROUND() function. It will round a value (whether calculated or loaded from a data source) to the number of decimals specified as the second parameter of the function-zero in this example.

ROUND( ) is only one of the functions that you can choose when truncating or rounding values. The DAX functions that carry out rounding and truncation are given in Table 4-3.

Table 4-3. *DAX Rounding and Truncation Functions*

Function	Description	Example
ROUND()	Rounds the value to 0 if the second parameter is zero. If the second parameter is greater than zero, the function rounds the value to the number of decimals indicated by the second parameter. If the second parameter is less than zero, the figure to the left of the decimal is rounded to the nearest 10 (for a second parameter of -1), 100 (for a second parameter of -2), etc.	ROUND([CostPrice], 2)
ROUNDDOWN()	Rounds the value down to 0 if the second parameter is zero. If the second parameter is greater than zero, the function rounds the value down to the number of decimals indicated by the second parameter. If the second parameter is less than zero, the figure to the left of the decimal is rounded down to the nearest 10 (for a second parameter of -1), 100 (for a second parameter of -2), etc. The value will always be rounded down, and never up	ROUNDDOWN([CostPrice], 2)

(continued)

Table 4-3. (*continued*)

Function	Description	Example
ROUNDUP()	Rounds the value up to 0 if the second parameter is zero. If the second parameter is greater than zero, the function rounds the value up to the number of decimals indicated by the second parameter. If the second parameter is less than zero, the figure to the left of the decimal is rounded up to the nearest 10 (for a second parameter of -1), 100 (for a second parameter of -2), etc. The value will always be rounded up, and never down	ROUNDUP([CostPrice], 2)
MROUND()	Rounds the value to the nearest multiple of the second parameter	MROUND(([CostPrice], 2)
TRUNC()	Removes the decimals from a value	TRUNC([CostPrice])
INT()	Rounds down (or up if the number is negative) to the nearest integer	INT([CostPrice])
FLOOR()	Rounds down to the nearest multiple of the second parameter	FLOOR([CostPrice], 2)
CEILING()	Rounds up to the nearest multiple of the second parameter	CEILING([CostPrice], .2)
FIXED()	Rounds a value to the number of decimals indicated by the second parameter and converts the result to a text	FIXED([CostPrice], 2)

Note   Remember that using these functions will modify data, whereas formatting numbers only changes their appearance.

Cascading Column Calculations

New calculated columns can refer to previously created calculated columns. This apparently anodyne phrase hides one of the most powerful features of Power BI Desktop - the ability to create spreadsheet - like links between columns where a change in one column ripples through the whole data model.

This implies that you will help yourself if you build the columns in a logical sequence so that you always proceed step by step and do not find yourself trying to create a calculation that requires a column that you have not created yet. Another really helpful aspect of new columns is that if you rename a column, Power BI Desktop will automatically update all formulas that used the previous column name, as well as using the new name in any visualizations that you have already created. This makes Power BI Desktop a truly pliant and forgiving tool to work with.

So if we take the DAX formulas that you have created so far, you have the Gross Margin that depends on the data for the sale price and the calculation of the Direct Costs - which itself is based on the data for the cost price, spare parts, and labor cost. If you are a spreadsheet user, you will probably not be surprised to see that any change to any of the source data for the four elements will cause both the Direct Costs and Gross Margin columns to be recalculated.

You can see this a little more graphically in Figure 4-12.

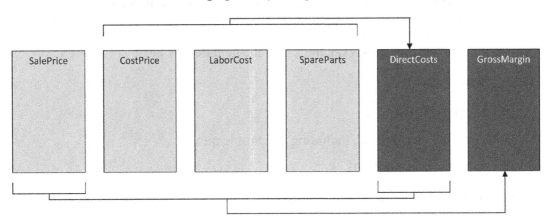

Figure 4-12. *Cascading column calculations*

Applying a Specific Format to a Calculation

Sometimes you will want to display a number in a particular way. You may need to do this to fit more information along the axis of a chart for instance. In cases like these you can, in effect, duplicate a column and reformat the data so that you can use it for specific visualizations. As an example of this - and to show how functions can be added to formulas that contain math - you will convert the cost of vehicles and any spare parts from pounds sterling to US dollars and then format the result in dollars.

1. In the Power BI Desktop window, make sure that you are in data view.

2. Click on the Sales table.

3. In the Table tools ribbon, click the New Column button. A new empty column will appear to the right of the final column of data. This column is currently entitled Column and is highlighted.

4. The formula bar above the table of data will display Column = .

5. In the Formula bar click to the right of the equals sign.

6. Click to the right of the equals sign.

7. Type **FORMAT(**. Once the function appears in the popup menu, you can select it if you prefer.

8. Enter a left square bracket and select the field [SalePrice].

9. Enter *** 1.6**.

10. Enter a comma.

11. Enter the text **"Fixed"**, including the double quotes.

12. Add a right parenthesis.

13. Still inside the Formula Bar, replace Column with **Invoice In Dollars**.

14. To the right of the equals sign enter **"$" &** (including the double quotes). The formula Bar will read

    ```
    Invoice In Dollars =
    "$ " & Format([SalePrice] * 1.6, "Fixed")
    ```

15. Press Enter (or click the tick icon in the Formula bar). The column
     will appear containing the same number as the one in the Cost
     Plus Spares column. However it will be formatted *as a text* and will
     be preceded by a dollar sign. Figure 4-13 shows a few records from
     this new column.

Figure 4-13. *Applying a custom numeric format*

The FORMAT( ) function can be applied equally well to dates and times as to
numeric columns - as you will see later in this chapter. Indeed it offers a wealth of
possibilities. So many, in fact, that rather than illustrate all of them Tables 4-4 and 4-5
contain the essential predefined formats that you can apply to columns of numbers.
Should you wish to create your own highly specific date and number formats, you can
assemble these using the format code elements given in these two tables.

Table 4-4. *Predefined Currency Formats*

Format Code	Description	Example		Comments
Currency	Currency	`FORMAT(Sales[CostPrice], "Currency"')`		The currency indicator will depend on the PC's settings and language used
Scientific	Exponential or scientific notation	`FORMAT(Sales[CostPrice], "Scientific")`		This is also called the scientific format
Fixed	Fixed number of decimals	`FORMAT(Sales[CostPrice],  "Fixed")`		Displays at least one figure to the left of the decimal (even if it is a zero) and two decimals
General Number	No format	`FORMAT(Sales[CostPrice],  "General Number"')`		Displays the number with no thousand separators
Percent	Percentage (and divided by 100)	`FORMAT(Sales[SalePriceToSalesCostsRatio],  "Percent"')`		Displays the number as a percentage

Table 4-5. *Custom Number Formats*

Format Code	Description	Comments
0	The zero placeholder	Adds a zero even if no number is present
#	The digit placeholder	Represents a number if one is present
.	The decimal character	Sets the character that will be used before the decimals
,	The thousands separator	Defines the thousands separator
%	Percentage symbol	Adds a Percentage symbol

Using the custom number formats is not difficult, but rather than explain all the permutations laboriously, here are a couple of examples to help you to see how they work.

- `FORMAT([Cost Plus Spares], "#,#.00")` will give you 44,500.00 (and any figure less than 1 will have a nothing to the left of the decimal)

- `FORMAT([Cost Plus Spares], "0.0")` will give you 12250.0 (and any figure less than 1 will have a 0 to the left of the decimal)

Note   Using the DAX FORMAT() function will convert the number to a text. This means that it can no longer be aggregated in visuals.

Calculation Options

I imagine that you have not had to worry about recalculation of Power BI Desktop workbooks if you have been using relatively small datasets like the sample data for this book. If, however, you are using vast amounts of data (and, after all, this is what Power BI Desktop was designed for), then recalculation could become a subject that you need to master.

By default Power BI Desktop will recalculate all calculated columns and measures when there is a change in the dataset. These are the main operations that can trigger a recalculation:

- Data from an external data source (of any kind) have been updated.

- Data from an external data source have been filtered.

- You have changed the name of a table or column.

- You have added, modified, or deleted relationships between tables.

- You have altered any formula for a calculated column or a measure.

- You have added new calculated columns or measures.

More generally, if you want to be sure that your data is up-to-date, you should probably update the data. You can do this by clicking the Refresh button in the Home ribbon.

Conclusion

This chapter introduced you to some of the core techniques that you can apply to extend a Power BI Desktop data model with further metrics. These additional elements were in the form of new columns that you added to many of the data tables that you had previously loaded, cleansed, and assembled into a structured data model.

All the added columns were based on DAX, the Power BI Desktop formula language. As you saw, this language is not especially difficult and is fairly close to the Excel formula language.

You also learned how to concatenate fields and how to perform basic arithmetic. You then discovered how to separate out date and time elements as new columns that can be used to visualize data in dashboards.

This brief introduction is nonetheless only a quick foretaste of the power of DAX. There is much, much more that can be accomplished to prepare the quantitative analyses that you are likely to need to produce telling visuals with Power BI Desktop. So it is time to move on to the next chapter and take a look at the next feature of DAX - adding new columns using data from multiple tables.

Should you wish to see the formulas that were created in this chapter, you can find them in the file PowerBIDimensionalWithColumns.pbix.

CHAPTER 5

Calculating Across Tables

In the previous chapter you started on the journey to extending a data model with calculated columns. This approach was limited to adding calculated columns that only referenced columns inside the current table. However it is possible in most cases to create a calculated column that uses the relationships between tables in the data model to create calculations that refer to columns in other tables. This is the next step in learning how to use DAX for calculated columns.

In this chapter you will build on the data model that includes the calculated columns that you saw in the last chapter. To save you having to build this file, it is available in the sample files as PrestigeCarsDimensionalWithColumns.Pbix.

Calculating Across Tables

If your data model is not complex - and particularly if it consists of a single table - then most calculations should be simple. All you will have to do is to follow the principle of building math expressions using column names and arithmetic operators and text functions.

The real world of data analysis is rarely this uncomplicated. In most cases you will have metrics on one table that you will need to apply in a calculation to figures in a completely different table. Power BI Desktop makes these "cross-table" calculations really easy *if* you have defined a coherent data model (and can be done even if tables are not joined in a data model, as you will discover in Chapter 17).

As an example let's see how to add the client and the country as a new column to the Sales table.

1. Open the sample file PrestigeCarsDimensionalWithColumns.Pbix.

2. In Data View select the *Sales* table.

3. Click New Column in the Table tools ribbon.

117

© Adam Aspin 2023

A. Aspin, *Pro DAX and Data Modeling in Power BI*, https://doi.org/10.1007/978-1-4842-8995-2_5

4. Replace the word Column in the formula bar with the word Client.

5. Start typing the keyword **RELATED**, and select this function once you have limited the selection of functions in the popup list (alternatively you can type the whole word and a left parenthesis).

6. Enter a single quote. The list of available fields that are accessible from across the data model will appear. You can see this in Figure 5-1.

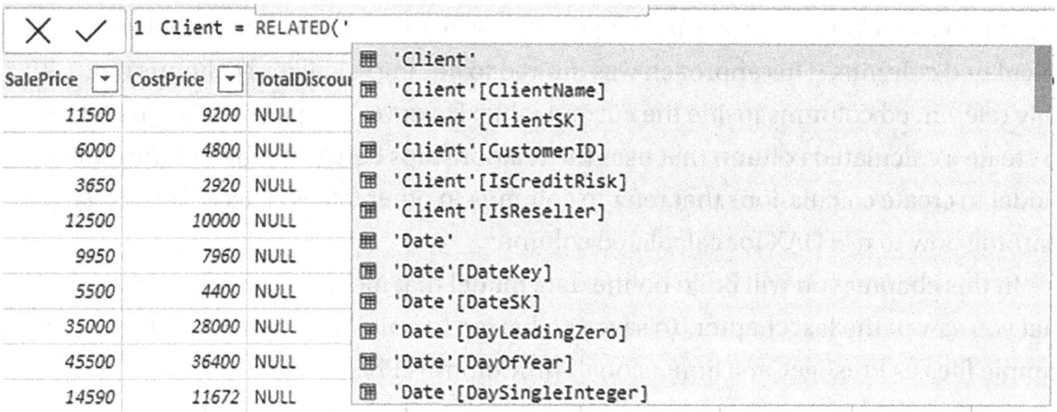

Figure 5-1. *Displaying available related fields*

7. Select the 'Client'[ClientName] field.

8. Enter a right parenthesis to close the RELATED() function.

9. Enter **& " - " &**.

10. Enter **RELATED(** and a single quote. The popup list will display all related tables that are accessible using the data model.

11. Select the field 'Geography'[CountryName].

12. Enter a right parenthesis. The Formula bar should contain the following formula:

```
Client =
RELATED('Client'[ClientName]) & " - " & RELATED('Geography
'[CountryName])
```

13. Click the check icon in the Formula Bar or press Enter to complete the definition of the calculated column.

You can now see a new column added to the right of the Sales table. You can see this in Figure 5-2.

SalePrice	CostPrice	TotalDiscount	DeliveryCharge	SpareParts	LaborCost	ClientSK	VehicleSK	GeographySK	DateSK	ID	Gross Profit	Client
£11,500.00	£9,200.00		£150.00	£750.00	£500.00	87	229	1	20190216	4	£2,300.00	WunderKar - Germany
£6,000.00	£4,800.00		£150.00	£750.00	£500.00	27	4	11	20200217	43	£1,200.00	Glitz - Germany
£3,650.00	£2,920.00		£150.00	£750.00	£500.00	46	181	20	20200228	45	£730.00	Magic Motors - United Kingdom
£12,500.00	£10,000.00		£150.00	£750.00	£500.00	86	11	31	20200430	51	£2,500.00	Wonderland Wheels - United Kingdom
£9,950.00	£7,960.00		£150.00	£750.00	£500.00	2	165	47	20200530	52	£1,990.00	Alexei Tolstoi - United Kingdom
£5,500.00	£4,400.00		£150.00	£750.00	£500.00	43	280	7	20200615	55	£1,100.00	M. Pierre Dubois - France
£35,000.00	£28,000.00		£550.00	£750.00	£500.00	59	166	25	20200615	57	£7,000.00	Peter Smith - United Kingdom
£45,500.00	£36,400.00		£550.00	£750.00	£500.00	84	42	7	20200615	59	£9,100.00	Vive La Vitesse - France
£14,590.00	£11,672.00		£150.00	£750.00	£500.00	63	286	8	20200726	69	£2,918.00	Prestissimo! - Italy
£16,500.00	£13,200.00		£150.00	£750.00	£500.00	58	229	61	20200813	79	£3,300.00	Peter McLuckie - United Kingdom
£19,500.00	£15,600.00		£150.00	£750.00	£500.00	27	248	11	20200819	85	£3,900.00	Glitz - Germany

Figure 5-2. *A calculated column with related fields*

This added column contains the client and the country of sale for every vehicle sold, even if the client name is in one table and the country is in a separate table. This is all thanks to the RELATED() function which links fields from different tables using the joins that you defined in the data model.

If you are an Excel user who has spent hours - or even days - wrestling with the Excel *Lookup()* function, then you are probably feeling an immense sense of relief. For it really is this easy to lookup values in another table in Power BI Desktop. Once again (and at risk of laboring the point), if you have a coherent data model, then you are building the foundations for simple and efficient data analysis further down the line using DAX.

Counting Reference Elements

Data models are often assembled to make the best use of reference elements. The PrestigeCarsDimensionalWithColumns data model has "lookup" tables (Client, Vehicle, and Geography) that contain essential information that you could need to analyze the underlying data. So as an example of how the data model can be put to good use, let's see how DAX can add a calculated column to the Geography table that displays the number of sales per country.

This challenge will introduce two new DAX elements:

- The *COUNTROWS()* function

- The *RELATEDTABLE()* function

As its name implies, the COUNTROWS() function will count a number of rows in a table. The RELATEDTABLE() function allows you to do an "aggregated" lookup from a table that contains a unique list of elements (such as countries) that is joined to a table that has many elements which are joined to the reference table. In essence, you will be adding a column to the "One" side of a join to aggregate elements from the "many" side of the join.

Once again, because the data model has been set up coherently, and the Geography table is joined to the Sales table, using the COUNTROWS() and RELATEDTABLE() functions together will not just return the number of records in a table, but will calculate the number of records for each element in the table where it is being applied. This means that any elements from the Geography table that exist in the Sales table can be identified because the Sales table is using the Geography table as a lookup table.

Here is an example of how you can use these two functions to count reference elements (still using the sample file PrestigeCarsDimensionalWithColumns.pbix):

1. In Data View select the *Client* table.

2. Click New Column in the Table tools ribbon.

3. Enter the formula:

    ```
    Sales Per Customer = COUNTROWS(RELATEDTABLE(Sales))
    ```

4. Click the tick icon in the Formula Bar (or press the Enter key).
    The number of customers for each country will appear in the new column named Sales Per Customer.

The Countries table now looks like it does in Figure 5-3.

ClientName	ClientSK	IsReseller	IsCreditRisk	CustomerID	ClientSince	ClientFlag	ClientWebsite	First Bought	Sales Per Customer
Alex McWhirter	1	0	False	85	31 October 2022		Http://www.calidra.co.uk	2022	1
Alicia Almodovar	3	0	False	39	30 October 2020		Http://www.calidra.co.uk	2020	7
Andrea Tarbuck	4	0	False	60	17 February 2022		Http://www.calidra.co.uk	2022	2
Andy Cheshire	5	0	False	87	31 October 2022		Http://www.calidra.co.uk	2022	2
Beltway Prestige Driving	8	1	False	61	20 April 2022		Http://www.calidra.co.uk	2022	2
Birmingham Executive Prestige Vehicles	9	1	False	3	03 February 2019		Http://www.calidra.co.uk	2019	6
Bling Bling S.A.	10	1	False	75	31 July 2022		Http://www.calidra.co.uk	2022	1
Boris Spry	12	0	False	59	17 February 2022		Http://www.calidra.co.uk	2022	3
Bravissimai	13	1	False	76	31 July 2022		Http://www.calidra.co.uk	2022	2
Capots Reluisants S.A.	14	1	False	46	23 May 2021		Http://www.calidra.co.uk	2021	6
Casseroles Chromes	15	1	False	5	02 January 2019		Http://www.calidra.co.uk	2019	4
Clubbing Cars	16	1	False	52	02 January 2022		Http://www.calidra.co.uk	2022	2
Diplomatic Cars	18	1	False	34	23 August 2020		Http://www.calidra.co.uk	2020	3
Eat My Exhaust Ltd	19	1	False	7	30 March 2019		Http://www.calidra.co.uk	2019	4
El Sport	20	1	False	74	31 July 2022		Http://www.calidra.co.uk	2022	1

Figure 5-3. *Using the Countrows() function to calculate the number of clients per country*

Using RELATED( ) to Traverse a Simple Data Model

So far in this chapter you have seen that you can add calculated columns that use the data model to allow you to

- Look up a value from a reference (or lookup) table

- Aggregate a series of values from a detail table inside a lookup table

However, these are fairly simple examples. The nature and structure of a Power BI Desktop data model control where you can add new columns from another table. For the moment it is best to accept that you can only bring data elements into one table using the RELATED() function if the data is reference (or "lookup") data from another table.

This means that tables like Geography, Client, or Vehicle cannot pull back data from another table using the RELATED() function. This is because they are "lookup" tables, and contain reference data that appears only once, but can be used many times in other tables.

However, you cannot pull data into the Clients table from Sales using the RELATED() function this simply. This is because when there are many sales for a client, Power BI Desktop does not know which row to select and return to the destination table.

In data modeling terms (and as you saw in Chapter 2), tables such as Geography, Client, or Vehicle are the "one" side of a relationship, whereas tables like Sales are on the "many" side of a relationship. In Power BI Desktop (as is the case in a relational database), you can only look up data from the "many" side.

So essentially the table where you add a new column has the potential to reach through most if not all of the data model and return data from many other tables. Providing, once again, that the data model has been constructed in a coherent manner.

Using RELATED( ) to Traverse a Complex Data Model

However, not all data models are star schemas like the model in the file PrestigeCarsDimensionalWithColumns.Pbix. If you refer back to Chapter 1 and the Power BI Desktop file PrestigeCarsRelational.Pbix that you can see in Figure 1-14, the structure is a little more complicated.

In a model such as this one, where there are many more "lateral" joins between tables (as opposed to the central focus of the joins in the star schema), it may be possible to look up data across several tables.

Suppose, for instance, that you start adding a calculated column to the Stock table in PrestigeCarsRelational.Pbix. The moment that you enter and select the RELATED() function, you see a large list of available fields that you can use as the basis for the new column. You can see this in Figure 5-4.

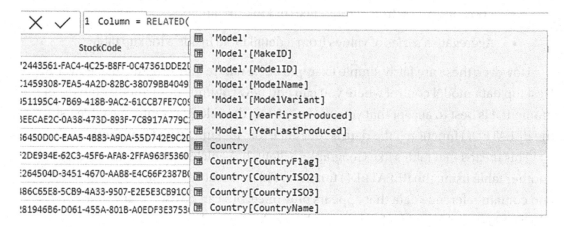

Figure 5-4. *The popup list of related fields in a complex data model*

In fact the Sales table in the PrestigeCarsRelational.Pbix file can pull in fields from the following tables:

- Make
- Model
- Country
- Customer
- Sales
- SalesDetails

However, the capacity to "see" fields from across the data model depends on how (and how well) the data model has been set up. Specifically it will depend on whether cross-filtering has been allowed. So this is what we need to look at next.

Cross Filter Direction

While a thoughtful initial data model means that you can reach across from many tables into many (or most) other tables to pick and choose any piece of data, this is not always the case. Indeed, it might not even be possible or desirable.

It may seem strange at first sight to have set up a data model whose main purpose is to join tables into a holistic set of data – only to be told that all the data may not be available to query in all circumstances. However, this apparent contradiction can easily be resolved. However, before giving technical solutions, it is important to understand a new concept. This concept is cross filter direction.

When establishing relationships in Chapter 1, you met the Create (and edit) relationship dialog. This dialog contained one element that was not explained previously. Figure 5-5 shows the cross filter direction popup in the Edit relationship dialog.

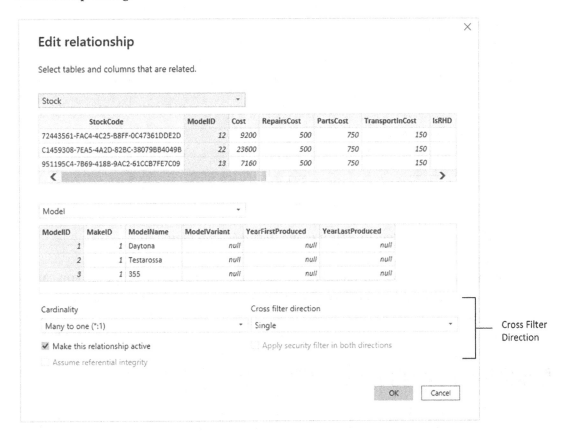

Figure 5-5. *The cross filter direction popup in the Edit relationship dialog*

At this juncture it is important to understand that when you create relationships (or if Power BI Desktop creates them automatically) a "one way system" is created inside the data model. What this does (to accelerate querying amongst other things) is to set table relationships to allow lookup tables to filter tables that contain details but not the other way round. In data modeling terms, the "one" side can filter the "many" side – but not the reverse. Alternatively, a bidirectional relationship is automatically created for one to one or many to many relationships.

This setting is immediately visible in the Model view, as you can see in Figure 5-6.

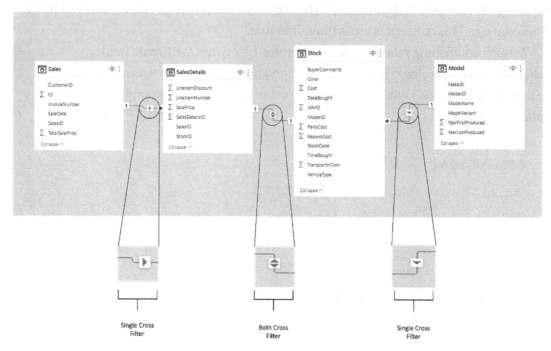

Figure 5-6. *The cross filter direction icons in the Model view*

In this figure (a subset of the tables in the file PrestigeCarsRelational.pbix), you can see that there are two single cross filter directions and one cross filter that allows filtering in both directions.

Modifying Cross Filter Direction

Cross filtering defines how filters propagate between tables. This generic explanation is probably easier to understand with a simple example.

1. Open - or return to - the PrestigeCarsDimensionalWithColumns.
    pbix file.

2. In Data View select the *Client* table.

3. Add a new (calculated) column using the formula

    ```
    Vehicles Per Customer = COUNTROWS(RELATEDTABLE('Vehicle'))
    ```

4. The table now looks like Figure 5-7 (only the top few records
    are shown).

ClientName	ClientSK	IsReseller	IsCreditRisk	CustomerID	ClientSince	StatusFlag	Website	Sales Per Customer
Alexei Tolstoi	2	0	False	15	15 July 2019		www.calidra.co.uk	300
Alicia Almodovar	3	0	False	39	30 October 2020		www.calidra.co.uk	300
Andrea Tarbuck	4	0	False	60	17 February 2022		www.calidra.co.uk	300
Andy Cheshire	5	0	False	87	31 October 2022		www.calidra.co.uk	300
Antonio Maura	6	0	False	48	15 June 2021		www.calidra.co.uk	300
Autos Sportivos	7	1	False	42	09 May 2021		www.calidra.co.uk	300

Formula bar: 1 Sales Per Customer = COUNTROWS(RELATEDTABLE('Vehicle'))

Figure 5-7. *Filter blocked by a cross filter setting*

Clearly this output is wrong. What is happening is that the total number of rows in
the Vehicle table is returned for each row in the Client table. This is because the cross
filter setting linking the Vehicle table to the rest of the data model is set to *Single*. This
means that the Vehicle table can filter other tables – but *cannot* be filtered by other
selections in the data model.

Note   You can think of the single triangle in the relationship between tables as
a "stop sign" that allows filtering in the direction of the top of the triangle and
prevents filtering in the table at the base of the triangle.

This way of functioning can be changed like this:

1. Switch to Model view.

2. Double-click on the relationship between the Vehicle and Sales
    tables. The Edit relationship dialog will appear (as seen above in
    Figure 5-6).

3. Select Both as the cross filter direction in the Cross filter direction popup.

4. Click OK. In the Model view the relationship between Vehicle and Sales is now a two-headed triangle.

5. Switch to Data view. The table now looks like the one in Figure 5-8.

| | 1 Sales Per Customer = COUNTROWS(RELATEDTABLE('Vehicle')) | | | | | | | | |

ClientName	ClientSK	IsReseller	IsCreditRisk	CustomerID	ClientSince	StatusFlag	Website	Sales Per Customer
Alexei Tolstoi	2	0	False	15	15 July 2019		www.calidra.co.uk	8
Alicia Almodovar	3	0	False	39	30 October 2020		www.calidra.co.uk	7
Andrea Tarbuck	4	0	False	60	17 February 2022		www.calidra.co.uk	2
Andy Cheshire	5	0	False	87	31 October 2022		www.calidra.co.uk	2
Antonio Maura	6	0	False	48	15 June 2021		www.calidra.co.uk	9
Autos Sportivos	7	1	False	42	09 May 2021		www.calidra.co.uk	1

Figure 5-8. *The cross filter setting allowing correct filtering*

The result of this operation is that filters set in the Client table are no longer blocked from being applied to the Vehicle table.

What this means in practice is that you may need to set some of the cross filter direction settings in your data model to allow cross filtering in both directions. This may slow down calculations, and there are advanced DAX alternatives to resetting cross filter direction. You will see this in Chapter 17. However, if you are starting a data model, it is probably more important to see correct results rather than aim for high-level technical perfection too soon.

Note   Bi-directional relationships can *introduce* inaccuracies in results if not implemented correctly, rather than fix them. This is because bi-directional relationships can cause ambiguous data models. So this technique should be applied sparingly.

There are a handful of points that need to be made concerning cross filtering:

- Cross filtering flows through a data model until a cross filter set to "single" is met. The filtering cannot progress to the table on the "one" side of the join or beyond to any other tables in the data model. In the data model for the file PrestigeCarsRelational.pbix, this means that blocking filtering on the Model table prevents the Make table (which

is joined to the Model table) being filtered by Customer or Country.
Equally, Customer and Country cannot be filtered by Make or Model
in the initial data model.

- One to one and many to many relationships are automatically set to
  "Both" by Power BI Desktop.

- Cross filtering also affects the output in tables. You can see how
  single cross filtering can result in surprising outputs in Figure 5-9
  (remember that the tables are aggregating models by customer
  whereas the Data view shows individual sales).

Cross Filter Direction: Both

CustomerName	Count of ModelName
	32
Alex McWhirter	1
Alexei Tolstoi	8
Alicia Almodovar	7
Andrea Tarbuck	2
Andy Cheshire	2
Total	**100**

Cross Filter Direction: Single

CustomerName	Count of ModelName
	100
Alex McWhirter	100
Alexei Tolstoi	100
Alicia Almodovar	100
Andrea Tarbuck	100
Andy Cheshire	100
Total	**100**

Figure 5-9. *Cross filtering tables*

Using Functions in New Columns

Over a few pages you have seen just how easy it is to extend a data model with some
essential metrics that you can then use in Power BI Desktop. Yet we have only performed
simple arithmetic to achieve our ends. Power BI Desktop can, of course, do much more
than just carry out basic arithmetic.

Statistical Functions in Calculated Columns

As you might expect in a piece of analytical software, Power BI Desktop is particularly
good at carrying out aggregations. Moreover, statistical functions (which is a more
refined way of defining aggregations in Power BI Desktop) can be applied anywhere in
the data model without needing out use the RELATED() or RELATEDTABLE() functions.

As a first step – and to show how statistical functions operate – let's suppose that you wish to see (and I realize that this is outrageously simplistic) the average mileage for all cars as a new column in the SalesInfo table.

1. Open the sample file PrestigeCarsDimensionalWithColumns.pbix.

2. Switch to Data view.

3. Select the *SalesInfo* table in the Fields pane.

4. In the Table tools ribbon click the New column button.

5. Type in (and/or select from the popup windows) the following formula

   ```
   Average Mileage = AVERAGE(SalesInfo[Mileage])
   ```

6. Confirm the formula by pressing Enter or clicking the tick icon in the Formula bar.

You will see the calculation of average mileage for all the vehicles in the newly created column (after scrolling down a few rows) as shown in Figure 5-10.

StockCode	Mileage	BuyerComments	ID	Mileage Range	DateBought	VehicleAgeInYears	Vehicle Age Category	Vehicle Age Category Sort	Special Sales	Average Mileage
72443561-FAC4-4C25-B8FF-0C47361DDE2D	33250		4	Low	14/02/2019	2.93832144216134	Under 5	1	Normal	49031.1350574713
21890556-D1C4-48C1-B0C8-40533195SE9D	33250		5	Low	01/01/2019	3.05886938736682	Under 5	1	Normal	49031.1350574713
6081DBE7-9AD6-4C64-A676-61D919E64979	33250		9	Low	04/04/2019	2.8040748668187	Under 5	1	Normal	49031.1350574713
F3A2712D-20CA-495D-9F6A-8A4CA195248D	33250		11	Low	15/04/2019	2.7789378805175	Under 5	1	Normal	49031.1350574713
4C57F13A-E21B-4AAC-9E9D-A218D4C691C6	33250		13	Low	01/05/2019	3.04791048325723	Under 5	1	Normal	49031.1350574713
98C1E31A-4258-4F78-95D4-2365167E6F3F	33250		14	Low	15/05/2019	2.6917460996955B	Under 5	1	Normal	49031.1350574713
951195C4-7B69-4168-9AC2-61CCB7FE7C09	33250		15	Low	26/05/2019	2.66160911339421	Under 5	1	Normal	49031.1350574713
91CF8133-EF19-4C92-BEFB-6A24FD85EF3A	33250		16	Low	03/06/2019	2.8835269261339	Under 5	1	Normal	49031.1350574713
94FF5451-471C-4F17-BE27-BA55D3ECF5DC	33250		18	Low	10/07/2019	2.29448582572298	Under 5	1	Normal	49031.1350574713

Figure 5-10. *Using aggregation functions*

As you can see – and this is the point I am making – any aggregation applies to the *entire field* that is being aggregated. At least, this is the initial state of play. As you progress with DAX, you will see that shaping the extent of calculations is a major part of using DAX. For the moment, however, you need to understand that this is the default behavior.

With this new column in place, you can now see how each vehicle's mileage compares to the average. One way to do this is to use the newly added column as a reference in another new column – like this.

1. In Data view, and with the SalesInfo table selected, click New Column.

2. Enter and confirm the following formula

```
Mileage difference from average = [Average Mileage] - [Mileage]
```

You will see the output for all the vehicles in the newly created column as shown in Figure 5-11.

StockCode	Mileage	BuyerComments	ID	Mileage Range	DateBought	VehicleAgeInYears	Vehicle Age Category	Vehicle Age Category Sort	Special Sales	Average Mileage	Mileage difference from average
D63C8CC9-0B19-4B9C-9C8E-6C6370812041	79327		10	Medium	02/04/2019	2.96573226249365	Under 5	1	Normal	49031.1350574713	-30295.8649425287
EDCCE4B1-5DAB-4E2E-8F0B-798431B41575	77996		12	Medium	10/04/2019	2.73285535016488	Under 5	1	Normal	49031.1350574713	-28964.8649425287
8CCB9C13-AEDA-4467-A014-48F0C7A0D6A4	54825		17	Medium	05/06/2019	2.71641171940005	Under 5	1	Normal	49031.1350574713	-5793.86494252874
CE0A56A6-8218-4F4C-A0E2-68F30C9E4AE6	110537		22	High	25/07/2019	2.49725911280872	Under 5	1	Normal	49031.1350574713	-61505.8649425287
BEECAE2C-0A38-473D-B95F-7C8917A779C2	54825		23	Medium	01/08/2019	3.03970486623338	Under 5	1	Normal	49031.1350574713	-5793.86494252874
C001858B-085D-4648-8F0D-802609964C921	110537		25	High	11/09/2019	2.10408842687721	Under 5	1	Normal	49031.1350574713	-61505.8649425287
A6FCB276-6311-4B3E-9C99-23F197953F1C	92754		27	Medium	30/09/2019	2.3136774679731	Under 5	1	Normal	49031.1350574713	-43722.8649425287
E264504D-3481-4870-AAB8-E4C66F2387B0	77996		34	Medium	23/12/2019	2.08954048167173	Under 5	1	Normal	49031.1350574713	-28964.8649425287
A926886C-FC26-4EB8-9976-2DF7EDC48E92	92754		35	Medium	24/12/2019	2.08080075564433	Under 5	1	Normal	49031.1350574713	-43722.8649425287

Figure 5-11. *Applying calculations referencing aggregation functions*

In the real world I would suggest that you try and minimize the use of calculated columns when you can. While breaking down DAX formulas into multiple steps spread over several columns can be practical when you are learning DAX or attempting new and possibly complex formulas, it is not advisable in "real" data models for the following reasons:

- Calculated columns take up space in memory. This means:

    – Slower data loading times

    – Taking up space in the Power BI service – where all the data used is competing for resources

- Calculated columns are only refreshed at data load.

So, alternatively, the two columns used previously could be replaced by a single column using the following formula:

```
Mileage difference from average = AVERAGE(SalesInfo[Mileage]) - [Mileage]
```

Which also makes the point that you can combine statistical functions and arithmetic functions when creating calculated columns.

As an intrinsic part of the Microsoft BI offering DAX can, of course, calculate a range of aggregates. After all, analyzing totals, averages, minima, and maxima (among others) is a core aspect of much business intelligence.

Once again if you are an Excel or Microsoft Access user, you are probably feeling quite at ease with this way of working. Even if you are not a spreadsheet or database expert, you must surely be feeling reassured that creating calculations that apply instantly to an entire column is truly easy.

Now that you have seen the basic principles, take a look at some of the more common available aggregation functions that are described in Table 5-1.

Table 5-1. *DAX statistical functions*

Function	Description	Example
AVERAGE()	Calculates the average (the arithmetic mean) of the values in a column. Any non-numeric values are ignored	AVERAGE([Mileage])
AVERAGEA()	Calculates the average (the arithmetic mean) of the values in a column. Empty text, non-numeric values and FALSE values count as 0. TRUE values count as 1	AVERAGEA([Mileage])
COUNT()	Counts the number of non-blank in a column	COUNT([Mileage])
COUNTA()	Counts the number of cells in a column that contain any values. Boolean values are also counted	COUNTA([Mileage])
COUNTBLANK()	Counts the number of blank cells in a column	COUNTBLANK([Mileage])
COUNTROWS()	Counts the number of rows in a table	COUNTROWS(Vehicle)
DISTINCTCOUNT()	Counts the number of unique values in a column	DISTINCTCOUNT([Vehicle])
MAX()	Returns the largest value in a column	MAX([Mileage])
MAXA()	Returns the largest value in a column. Dates and logical values are also included	MAXA([Mileage])
MEDIAN()	Returns the median numeric value in a column	MEDIAN([Mileage])
MIN()	Returns the smallest value in a column	MIN([Mileage])
MINA()	Returns the smallest value in a column. Dates and logical values are also included	MINA([Mileage])

There are many more statistical functions in DAX, and you can take a deeper look at them in the Power BI online documentation. However for the moment the intention is not to blind you with science, but to introduce you more gently to the amazing power of DAX. For the moment, then, rest reassured that all your favorite Excel functions are present when it comes to calculating aggregate values in Power BI Desktop.

Note   You do not *have* to add the table name when applying a statistical function to a field. IF the field name is unique in the data model, then you can use the field name on its own. However it is considered best practice to indicate the table – and this can help you when debugging formulas at a later date.

Summarizing for Each Row in a Table

You saw previously how statistical functions can aggregate all the data in a column. One further aspect of calculated columns is returning the aggregate data for *each record* in the column.

In this case, suppose that you know that you can calculate the total number of sales (using the formula `Number of Sales = Count(Sales[ClientSK])`. However, what you want to do is to see the number of sales for each record in the Geography table (which corresponds to the sales per postcode as the postcode is the lowest level of detail in the Geography table). To do this (still in the file PrestigeCarsDimensionalWithColumns.pbix)

1. Switch to Data view and select the *Geography* table in the Fields pane.

2. In the Table tools ribbon click the New column button.

3. Type in (and/or select from the popup windows) the following formula:

   `Number of Sales Per Postcode = CALCULATE(COUNT(Sales[ClientSK]))`

4. Confirm the formula by pressing Enter or clicking the tick icon in the Formula bar.

You should see a new column at the right of the Geography table that looks like the one you can see in Figure 5-12.

✕ ✓	1 Number of Sales Per Postcode = CALCULATE(COUNT(Sales[ClientSK]))				

CountryName ▼	GeographySK ▼	Town ▼	PostCode ▼	SalesRegion ▼	Number of Sales Per Postcode ▼
Germany	1	Berlin		EMEA	4
Belgium	2	Brussels		EMEA	10
Switzerland	3	Lausanne		EMEA	7
United States	4	Los Angeles		North America	17
France	5	Lyon		EMEA	6
Spain	6	Madrid		EMEA	11
France	7	Marseille		EMEA	32
Italy	8	Milan		EMEA	11
United States	9	New York		North America	6
France	10	Paris		EMEA	31
Germany	11	Stuttgart		EMEA	9
United States	12	Washington		North America	8
Italy	13	Rome	120	EMEA	2
Italy	14	Rome	129	EMEA	1
Italy	15	Rome	175	EMEA	2

Figure 5-12. *Using CALCULATE() to aggregate per row*

This example is a very gentle introduction to the CALCULATE() function on DAX. This function is so far-reaching and powerful that it will be covered, progressively, in several chapters. For the moment it is worth noting that CALCULATE() adds a level of depth and complexity to any DAX function. When used in calculated columns, it *forces the calculation to be carried out at the level of the individual row.*

This is only a first step towards using CALCULATE(). You will be seeing a lot more of this function in chapters to come.

Limitations of Calculated Columns

While calculated columns are a boon when it comes to adding metrics that you need in your dashboards, you need to be aware of some of the limitations of calculated columns. These limitations may not be important to you - or be particularly relevant if you are dealing with small datasets. Yet it is probably best to be aware of the overall reasons why you may, in many cases, prefer to use other solutions (such as adding columns in Power Query or using measures) wherever possible.

- Calculated columns are recalculated when data is loaded or refreshed – or a new calculated column is added. This can slow down a data load.

- Calculated columns take up space in memory (as do columns added using Power Query). However calculated columns may not be as efficiently compressed in memory as columns created with Power Query - and measures take up no space at all.

- Filtering values in a calculated column does not cause calculated columns to be refreshed.

- Creating multiple intermediate calculated columns that "cascade" down to a final result will take up more space (and slow down the load process) more than a single, more complex, calculated column that delivers the same result.

Having said all this, calculated columns may be perfectly suited to your requirements and are certainly a good way to start learning DAX.

Conclusion

In this chapter you learned that calculated columns are not limited to using source fields from the table where you create the column. Two fundamental DAX functions-RELATED() and RELATEDTABLE() allow you to refer to fields in other tables in the data model when creating new columns. The actual approach that you take depends on whether you are looking up a single value or aggregating data from the other table.

The new columns created in this chapter are available in the Power BI Desktop file PrestigeCarsDimensionalWithColumnsAcrossTables.pbix. This file is, of course, available with the sample data for this book.

CHAPTER 6

DAX Logical Functions

Analytics frequently involves applying some core logic to data. So it will probably come as no surprise to learn that DAX (rather like Excel) contains a set of logical functions that you can use to add indicators to data as well as grouping and classifying data.

This chapter will be a gentle introduction to a few of the core DAX logic functions along with some ideas as to ways they can be used. For the moment you will only be looking at logic functions when used in calculated columns. From Chapter 8 onwards you will see how to build on this knowledge to apply logic to measures as well.

To continue building on the knowledge acquired in the previous chapters, this chapter will use the data that you have already extended previously. The source file to use is PrestigeCarsDimensionalWithColumnsAcrossTables.Pbix.

Simple Logic-the IF() Function

Having data available is always a necessary prerequisite for analysis. However the raw data that you have loaded may not always lend itself to being used in dashboard visualizations in an ideal way.

DAX can help you to see "the wood for the trees" in the thicket of data that underlies your data model. Let's begin by taking a look at a series of practical examples that extend your data in ways that use the capabilities of DAX to do the heavy lifting and let you focus on items that need your attention. The core enabler underpinning this is the IF() function.

© Adam Aspin 2023
A. Aspin, *Pro DAX and Data Modeling in Power BI*, https://doi.org/10.1007/978-1-4842-8995-2_6

Exception Indicators

As a first example of how to use the IF() function, suppose that you want to highlight any records where the cost of spare parts is over £2.000.00. This means comparing the contents of the column PartsCost to a fixed value (2000 in this example). If this test turns out to be true (i.e., the parts cost is over the threshold that you have set), then you want to display the words *Too High!* In a new column.

Here is how to add a column that applies this test to the data.

1. Open the PrestigeCarsDimensionalWithColumnsAcrossTables. pbix file.

2. Click on the Data View icon, and then click on the *Sales* table in the Fields list.

3. Click the New Column button in the Table tools ribbon. A new column named Column will appear at the right of any existing columns.

4. On the right of the equals sign, enter **IF(**. You will see that as you enter the first few characters, the list of functions will list all available functions beginning with these characters.

5. Press the **[** key. The list of available fields will appear.

6. Scroll down through the list of fields, and click on the field SpareParts.

7. Enter the greater than symbol - **>**.

8. Enter **2000**.

9. Enter a comma.

10. Enter the following text (including the double quotes) **"Too High!"**.

11. Enter a closing parenthesis - **)**. The code in the Formula Bar will look like this:

```
Column = IF([SpareParts]>2000,"Too High!")
```

12. Press Enter or click the tick icon in the Formula bar. The new
     column will display Too High for any rows where the cost of spares
     was over £2,000.00.

13. Rename the column **Excessive Parts Cost**.

If you scroll down the Sales table, you will start to see the flag "Too High!" in the
Excessive Parts Cost column. If you verify the SpareParts column for all rows that are
flagged, you will see that the value is over 2,000. You can see an example of the output in
Figure 6-1 in the next section.

Explaining the IF() Function

The IF() function can take up to three arguments or parameters (as the separate
elements that you enter between the parentheses are called). The first two are
compulsory; they are

- A *test* - in this case comparing the contents of a column to a
  fixed value

- The outcome if the test is *positive* (or "TRUE" in programming terms)

This was one simple test that helps you to isolate certain records. Later, when
building dashboards, you can use the contents of this new column as the basis for tables,
charts, and indeed just about any Power BI Desktop visualization.

The IF() function can also have a *third* argument - although this is optional. Let's
see an example of this in action as well. The third argument essentially means that the
outcome if the test is *negative* (or "FALSE" in programming terms). So let's presume that
you want to add a column that indicates whether a sale is exceptional or standard. As
you are now used to the approach that you need to take to enter calculated columns, I
will not spell out every step.

1. In the PrestigeCarsDimensionalWithColumnsAcrossTables.pbix
    file, click on the Data View icon, and then click on the Sales table
    in the Fields list.

2. Add a calculated column containing the following DAX

```
Super Sale = IF([SalePrice]>150000,"Brilliant", "Banal")
```

Once again, if you scroll down the table, you will see a few rows flagged as "Brilliant." These correspond to a sale price greater than 150,000. You can see this in Figure 6-1.

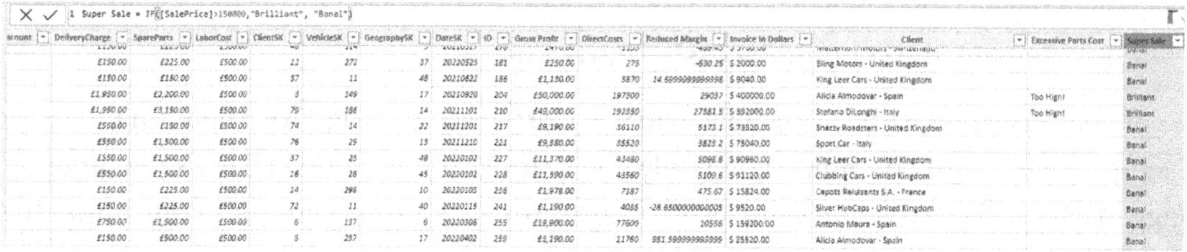

Figure 6-1. *A simple IF() function to aply a test to data*

Creating Alerts

When using the IF() function, the major focus is nearly always on the first argument - the test. After all, this is where you can apply the real force of DAX logic. So here is another example of an IF() function being used; only this time it is to create an alert based on slightly more complex logic. This time the objective is to detect records where the selling price of the vehicle is less than half the average sale price for all cars - using the AVERAGE() function that you learned about in the previous chapter.

1. In the PrestigeCarsDimensionalWithColumnsAcrossTables.pbix file, click on the Data View icon, and then click on the *Sales* table in the Fields list.

2. Click the New Column button in the Table tools ribbon. A new column named Column will appear at the right of any existing columns.

3. Enter the following DAX in the Formula Bar:

```
PriceCheck =
IF([CostPrice] >= AVERAGE([CostPrice]) *2,
"Cost Price too high",
"Cost Price OK"
    )
```

4. Press Enter or click the tick icon in the Formula bar. The new column will display Price too High or Cost Price OK depending on whether the cost price is more than half the average cost price or not.

You could then, for instance, use the results of the cost price test as the basis for a visualization to compare the expensive purchases with the others. You can see the Sales table with this new column in Figure 6-2.

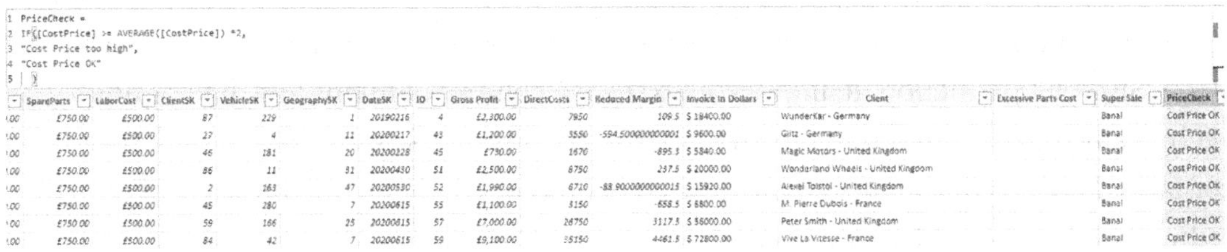

Figure 6-2. *An IF() function using a calculation to apply a test to data*

Comparison Operators

When carrying out tests like this, you will need to compare values. You may be used to the standard comparison operators that many programs and languages use (such as Excel), but for the sake of completeness Table 6-1 provides a list of the most frequently used operators that you could need.

Table 6-1. *DAX Comparison Operators*

Operator	Description
=	Equals
	Not equals to
<	Less than
>	Greater than
<=	Less than or equals to
>=	Greater than or equals to

Testing the Absence of Data

Let's see a practical example of how you might use an IF() function to validate data. Imagine that Prestige Cars is embarking on a "know your customer" program and you envisage a chart that compares the clients that have reliable postcodes with those that do not. This way you can make a business case for cleansing the data and potentially rooting out certain clients.

For the moment the Clients table either has or does not have a postcode (Zip code) for each customer. What you want is a clear extra column that contains either "HasPostCode" or "NoPostCode" to indicate whether there is a postcode present.

In this example you will not be testing numeric values, but will be looking at whether a record contains a value for a row. This means introducing a new DAX function. This is the ISBLANK() function. It allows you to see if a column contains any data or not. Technically this DAX function returns TRUE if the column is empty, and FALSE if it contains data. So you can nest it inside an IF() function to detect the *presence* of data, rather than looking at the data itself.

Here is how to create a clear indicator of the presence or absence of a postcode:

1. In the PrestigeCarsDimensionalWithColumnsAcrossTables.pbix file, click on the *Geography* table in the Fields list.

2. Click the New Column button in the Table tools ribbon.

3. Enter the following piece of DAX:

    ```
    HasPostCode = IF(ISBLANK([PostCode]),"NoPostCode",
    "HasPostCode")
    ```

4. Press Enter or click the tick icon in the Formula bar. The new column will display either NoPostCode or HasPostCode for every Client.

You can now use this new field to filter data in dashboards or in tables and charts to separate the clients that do or do not have postcodes. You can see the output from this in Figure 6-3.

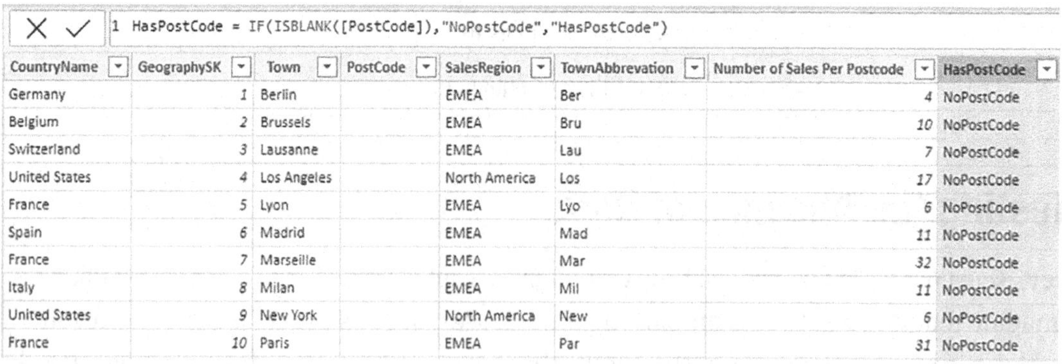

	1	HasPostCode = IF(ISBLANK([PostCode]),"NoPostCode","HasPostCode")						

CountryName	GeographySK	Town	PostCode	SalesRegion	TownAbbrevation	Number of Sales Per Postcode	HasPostCode
Germany	1	Berlin		EMEA	Ber	4	NoPostCode
Belgium	2	Brussels		EMEA	Bru	10	NoPostCode
Switzerland	3	Lausanne		EMEA	Lau	7	NoPostCode
United States	4	Los Angeles		North America	Los	17	NoPostCode
France	5	Lyon		EMEA	Lyo	6	NoPostCode
Spain	6	Madrid		EMEA	Mad	11	NoPostCode
France	7	Marseille		EMEA	Mar	32	NoPostCode
Italy	8	Milan		EMEA	Mil	11	NoPostCode
United States	9	New York		North America	New	6	NoPostCode
France	10	Paris		EMEA	Par	31	NoPostCode

Figure 6-3. Detecting empty cells using an IF() function

Nested IF() Functions

A frequent requirement in data analysis is to categorize records by ranges of values. Suppose, for instance, that you want to break down the sales of cars into low-, medium-, and high-mileage models. This requires more than a simple IF() function. However it is not very difficult, as all that is needed is to "nest" one IF() function inside another, thereby extending the test that is applied to cover three possible outcomes.

Here, then, is how to create a simple nested IF() function.

1. In the PrestigeCarsDimensionalWithColumnsAcrossTables.pbix file, click on the *SalesInfo* table in the Fields list, and click the New Column button in the Table tools ribbon.

2. Enter the following DAX snippet in the Formula bar:

```
Mileage Range =
IF([Mileage] <= 50000, "Low",
   IF([Mileage] < 100000, "Medium","High")
   )
```

3. Press Enter or click the tick icon in the Formula bar. The new column will display either Low, Medium, or High for every vehicle in the new Mileage Range column.

You have now categorized all the cars in SalesInfo by their mileage and can use the category flag in the Mileage Range column to create, for instance, a chart that shows the number of vehicles corresponding to each mileage category. Figure 6-4 shows some of the output from this new column (after scrolling down a few rows).

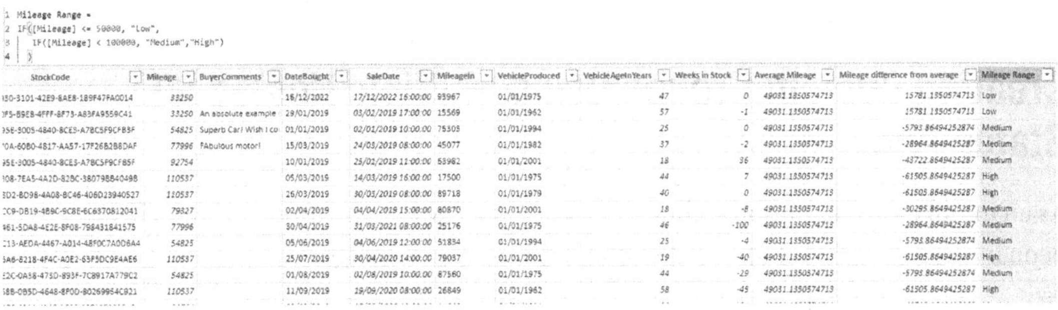

Figure 6-4. *Nested IF() functions*

A nested IF() function works like this:

- You set up a *first* test - in this example it is to flag all cars that have less than 50000 miles "on the clock."

- You specify what the outcome is if this initial test is positive. In this example the word *Low* will appear in the column.

- You then add a *second* test. By definition this will *only* apply to cars that have travelled more than 50000 miles - because otherwise the formula returned the word *Low*. So you add a higher threshold for the second test - 100000 miles in this example.

- If the record passes the test and the vehicle has travelled less than 100000 miles, then the word *Medium* will appear in the column. In all other cases (i.e., for all mileage over 100000 miles), the word *High* will be displayed in the column.

When writing nested IF() statements, the essential trick is to use a sequence of tests that follow a logical order from lowest to highest (or in some cases from highest to lowest). This way the succession of IF() statements acts like a series of hoops that catch the values and return an appropriate result.

You can nest up to 64 IF() statements in a single DAX expression. In fact, you can nest a maximum of 64 DAX expressions, whatever they are. However, to be realistic, nesting more than half a dozen IF() statements can be painful to write correctly, and getting the correct number of right parentheses in place can be tricky in practice. Nevertheless there may be many occasions when you will need to segment your data ready for your visualizations and dashboards even if it does mean grappling with complex nested IF() statements. So let's take a look at one of these to whet your appetite.

Creating Custom Groups Using Multiple Nested IF() Statements

As an initial example, suppose that you want to categorize the time that vehicles have been in stock. The CEO has decided that she wants four categories:

- 1 – for less than 0 weeks (just in case salespeople have been jumping the gun and over-anticipating stock levels)

- 2 – 1 or 2 weeks

- 3 – between 3 and 10 weeks

- 4 – over 10 weeks

The DAX to do this (added to the SalesInfo table) is the following. It requires four IF() functions.

```
Stock Category Classification =
IF([Weeks in Stock] < 0 , 1,
   IF(AND([Weeks in Stock] >= 0, [Weeks in Stock] <= 2), 2,
     IF(AND([Weeks in Stock] >= 3, [Weeks in Stock] <= 10), 3,
      IF([Weeks in Stock] >= 11, 4
      )
     )
    )
   )
```

To give you another example of a slightly more complex DAX function, but one that can be very necessary, consider the following requirement. Our data now has the car age, but we want to group the cars by age segments (or buckets if you prefer). So we will use a nested IF() function to do this. Then, to allow us to sort the column in a more coherent way, we will create a Sort By column for the new Vehicle Age Category column that we are creating. If you need to revise the concept of Sort by columns, we saw how to create and use them in Chapter 2.

In this example it is important to concentrate on the logic itself and explaining how complex IF() statements can be built.

A more deeply nested example is calculating the Vehicle Age Category in the SalesInfo table using the following code. This DAX snippet contains 6 nested IF() statements!

```
Vehicle Age Category=
IF([VehicleAgeInYears] <=5,"Under 5",
IF(AND([VehicleAgeInYears]>=6,[VehicleAgeInYears]<=10),
"7-10",
IF(AND([VehicleAgeInYears]>= 11,[VehicleAgeInYears]<=15),
"7-15",
IF(AND([VehicleAgeInYears]>=16,[VehicleAgeInYears]<=20),
"17-20",
```

```
IF(AND([VehicleAgeInYears]>=21,[VehicleAgeInYears]<=25),
"21-25",
IF(AND([VehicleAgeInYears]>=26,[VehicleAgeInYears]<=30),
"27-30", ">30"
                )
              )
            )
          )
        )
      )
```

The only slight problem with a great technique for segmenting data is that if you sort the Vehicle Age Category column, you will find that the category that corresponds to the highest age will appear at the top of the list. So you will need to add a *second* column that can be used as a sort order column for the new column that you just created. The code for this Vehicle Age Category Sort column (which should also be added to the SalesInfo table) is

```
Vehicle Age Category Sort=
IF([VehicleAgeInYears]<=5, "1",
    IF(AND([VehicleAgeInYears]>=6,           [VehicleAgeInYears]<=10),"2",
      IF(AND([VehicleAgeInYears]>= 11, [VehicleAgeInYears]<=15), "3",
        IF(AND([VehicleAgeInYears]>=16, [VehicleAgeInYears]<=20), "4" ,
          IF(AND([VehicleAgeInYears]>=21 , [VehicleAgeInYears]<=25), "5",
            IF(AND([VehicleAgeInYears]>=26, [VehicleAgeInYears]<=30),"6","7"
              )
            )
          )
        )
      )
    )
```

Figure 6-5 shows some of the output from these three new columns.

Stock Category Classification	▼	Vehicle Age Category	▼	Vehicle Age Category Sort	▼
1		17-20		4	
1		>30		7	
1		>30		7	
1		>30		7	
3		27-30		6	
3		>30		7	
3		>30		7	
4		27-30		6	
4		>30		7	
2		27-30		6	
1		>30		7	
1		>30		7	
1		17-20		4	
2		17-20		4	

Figure 6-5. *Complex nested IF() functions*

Although you can set the Sort by column in Model view, it can be faster (when dealing with ad-hoc requirements such as this one) simply to click inside the column to be sorted (Vehicle Age Category in this example) and in the Column tools ribbon click on the Sort by column button and select the appropriate column (Vehicle Age Category Sort in this example).

These formulas could have come straight from an Excel spreadsheet. Indeed, some 80 of the DAX functions are virtually identical to their Excel cousins - and over 200 have equivalents. So experience and imagination combined will show you that you have many ways to extend the data you imported by adding calculated columns. Even better, all calculated columns are updated when you refresh the data from the source. The only major caveat is that you have to be careful if you are tweaking the data connection **not** to delete any source columns on which a calculated column depends, or you will get errors in the Power BI Desktop table.

Tip   You could have created the formula in this example without creating the VehicleAge column first, as you could have used the formula that calculates the age of the car each time that you need the vehicle age. However, as you can imagine, it is easier to create a column that contains the vehicle age first and then refer to this in the Vehicle Age Category formula. This makes the more complex formula easier to read and also helps you to break down the analytical requirement into successive steps – which is good DAX development practice. Moreover you can always hide any "Intermediate" columns if you do not need them in dashboards and reports.

Multiline Formulas

All formulas that you create in Power BI Desktop will, by default, be on a single line that overflows onto the next line only when there is no more room in the Formula Bar. This can become an extremely tedious way of working, so it is worth knowing that you can tweak long formulas to force them to display over more than one line. All you have to do is force a line return inside the formula bar by pressing Shift-Enter where you want to force a new line. My experience is that Power BI Desktop will not let you create line breaks everywhere in a formula. Nonetheless, with a bit of trial and error, a more complicated formula, such as the Vehicle Age Category column that you created previously, can look like the formula in Figure 6-5.

Just in case you were wondering, you do **not** have to write formulas over multiple lines as I did just above. Indeed, the two formulas used to create complex nested IF() statements could be written as follows:

```
Vehicle Age Category=IF([VehicleAgeInYears] <=5,"Under 5",IF(AND([Vehicl
eAgeInYears]>=6,[VehicleAgeInYears]<=10),"4-10",IF(AND([VehicleAgeInYea
rs]>= 11,[VehicleAgeInYears]<=15),"11-15",IF(AND([VehicleAgeInYears]>=16
,[VehicleAgeInYears]<=20),"14-20",IF(AND([VehicleAgeInYears]>=21,[Vehicle
AgeInYears]<=25),"21-25",IF(AND([VehicleAgeInYears]>=26,[VehicleAgeInYea
rs]<=30),"24-30","Over 30"))))))
```

Or even

```
Vehicle Age Category Sort=IF([VehicleAgeInYears] <=5,"1",IF(AND([VehicleAge
InYears]>=6,[VehicleAgeInYears]<=10),"2",IF(AND([VehicleAgeInYears]>= 11,
[VehicleAgeInYears]<=15),"3",IF(AND([VehicleAgeInYears]>=16,
[VehicleAgeInYears]<=20),"4",IF(AND([VehicleAgeInYears]>=21,
[VehicleAgeInYears]<=25),"5",IF(AND([VehicleAgeInYears]>=26,
[VehicleAgeInYears]<=30),"6","7")))))))
```

I chose to write the formulas over multiple lines hoping that by doing so, I'd make the nested logic clearer. You can write your formulas in any way that suits you and that does not cause Power BI Desktop a problem.

In any case I recommend that you look at a fabulous tool that can format your DAX automatically. This is Dax Formatter. You can find it at www.daxformatter.com.

Making Good Use of the Formula Bar

If you only ever enter really simple formulas, then not only will you be extremely lucky, but you can content yourself with a single line in the formula bar. I doubt, however, that this is likely to be the case, as hopefully you will want to do great things with Power BI Desktop. It follows that you may soon be tired of creating long and complex DAX formulas in a limited space. So here is how to expand the formula bar — pretty much as you would in Excel.

1. Click the Expand icon at the right of the formula bar (the downward-facing chevron).

    The formula bar will increase in height to allow you to type and see several lines of text. To reduce the height of the formula bar and reset it to a single line, just click the Reduce icon at the right of the formula bar (which has now become an upward-facing chevron). You'll see this icon in Figure 6-6 momentarily.

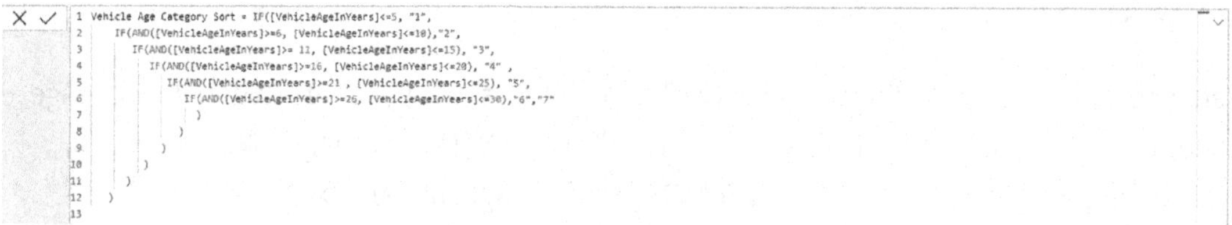

```
1  Vehicle Age Category Sort = IF([VehicleAgeInYears]<=5, "1",
2      IF(AND([VehicleAgeInYears]>=6, [VehicleAgeInYears]<=10),"2",
3          IF(AND([VehicleAgeInYears]>= 11, [VehicleAgeInYears]<=15), "3",
4              IF(AND([VehicleAgeInYears]>=16, [VehicleAgeInYears]<=20), "4" ,
5                  IF(AND([VehicleAgeInYears]>=21 , [VehicleAgeInYears]<=25), "5",
6                      IF(AND([VehicleAgeInYears]>=26, [VehicleAgeInYears]<=30),"6","7"
7                      )
8                  )
9              )
10          )
11      )
12  )
13
```

Figure 6-6. *Multi-line formulas*

Note   In the chapters that explain column-based calculations, I frequently emphasize that you need to prepare your metrics *before* building visualizations for dashboards and reports. In practice Power BI Desktop is extremely forgiving and immensely supple. It lets you switch from Report View to Data view at any time so that you can add any new columns or missing metrics. Indeed you can even add measures from the Fields list in Report View. However it can be more constructive to think through all your data requirements before rushing into the fun part that is creating dashboards. This approach can save you creating duplicate calculated columns and can help you to adopt a clear and coherent approach to naming new columns as well.

Keyboard Shortcuts in the Formula Bar

The Formula bar can prove to be a fairly efficient text editor to use when creating or modifying complex DAX formulas. The trick is to have an understanding of a range of useful keyboard shortcuts that you can apply when working.

The full list of available keyboard shortcuts is given in Table 6-2.

Table 6-2. *Keyboard Shortcuts in the Formula Bar*

Shortcut	Description
Ctrl+Shift+\	Jump to matching bracket
Ctrl+] / [	Indent/outdent line
Ctrl+F2	Select all occurrences of current word
Shift+Alt+→	Expand selection
Shift+Alt+←	Shrink selection
Home / End	Go to beginning/end of line
Ctrl+C	Copy line (empty selection)
Ctrl+G	Go to line…
Ctrl+M	Toggle Tab moves focus
Ctrl+U	Jumps to previous cursor position
Ctrl+X	Cut line (empty selection)
Shift+Enter	Insert line below
Ctrl+Shift+Enter	Insert line above
Ctrl+Shift+K	Delete line
Ctrl+Home	Go to beginning of the formula
Ctrl+End	Go to end of the formula
Ctrl+Shift+Alt + (arrow key)	Column (box) selection
Ctrl+Shift+Alt +PgUp/PgDn	Column (box) selection page up/down
Ctrl+Shift+L	Select all occurrences of the selected element in the formula
Ctrl+Alt+ ↑ / ↓	Insert cursor above / below
Shift+Alt + (drag mouse)	Column (box) selection
Shift+Alt + ↓ / ↑	Copy line up/down
Shift+Alt+I	Insert cursor at end of each line selected
Alt+ ↑ / ↓	Move line up/down
Alt+Click	Insert cursor

Complex Logic

Categorizing data can sometimes involve applying logic that is more complex than a single simple comparison. You could need to apply two or more conditions when evaluating a record, and this more intricate logic could require you to test the contents of more than one column.

Once again DAX can help you in circumstances like these. To explain by example, consider the following analytical challenge. You want to flag any vehicle that is a red or blue coupé. This example will show the basics of applying complex logic to data analysis with DAX.

1. In the PrestigeCarsDimensionalWithColumnsAcrossTables.pbix file, click on the *Vehicle* table in the Fields list, and click the New Column button in the Table tools ribbon.

2. Enter the formula:

```
Special Sales =
IF(
AND([VehicleType] = "Coupe", OR([Color] = " Black ", [Color] =
"Blue")),"Special","Normal")
```

3. Press Enter or click the tick icon in the Formula bar. The new column will display either Special or Normal for every vehicle in the new Special Sales column.

I realize that a formula like this can seem daunting at first sight. So let's take another look at this DAX expression formatted a little differently.

```
Special Sales = IF(
                AND(
                    [VehicleType] = "Coupe",
                    OR([Color] = "Black",
                       [Color] = "Blue"
                      )
                   )
                 ,"Special"
                 ,"Normal"
                 )
```

You can see the output from this calculated column in Figure 6-7.

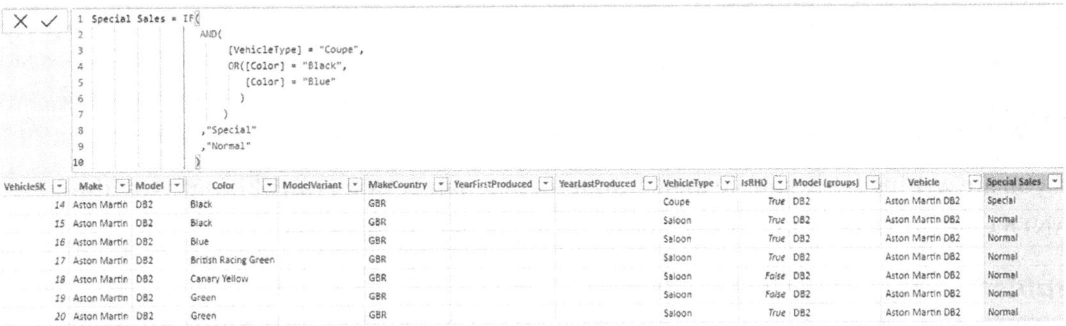

```
  1  Special Sales = IF(
  2                     AND(
  3                         [VehicleType] = "Coupe",
  4                         OR([Color] = "Black",
  5                            [Color] = "Blue"
  6                         )
  7                     )
  8                     ,"Special"
  9                     ,"Normal"
 10                  )
```

VehicleSK	Make	Model	Color	ModelVariant	MakeCountry	YearFirstProduced	YearLastProduced	VehicleType	IsRHD	Model (groups)	Vehicle	Special Sales
14	Aston Martin	DB2	Black		GBR			Coupe	True	DB2	Aston Martin DB2	Special
15	Aston Martin	DB2	Black		GBR			Saloon	True	DB2	Aston Martin DB2	Normal
16	Aston Martin	DB2	Blue		GBR			Saloon	True	DB2	Aston Martin DB2	Normal
17	Aston Martin	DB2	British Racing Green		GBR			Saloon	True	DB2	Aston Martin DB2	Normal
18	Aston Martin	DB2	Canary Yellow		GBR			Saloon	False	DB2	Aston Martin DB2	Normal
19	Aston Martin	DB2	Green		GBR			Saloon	False	DB2	Aston Martin DB2	Normal
20	Aston Martin	DB2	Green		GBR			Saloon	True	DB2	Aston Martin DB2	Normal

Figure 6-7. *Complex IF() logic*

As you can see, the expression is at its heart an IF() expression. As such remember that it consists of three parts:

- A *test* (The car is a coupé that is either black or blue)

- An *outcome for a positive result* (display "Special")

- An *outcome for a negative result* (display "Normal")

Possibly the only tricky bit now is the test itself. As it is built on more complex logic, it requires a little explanation.

- *Firstly* you have stated that the test is in several parts, all of which must be true for a record to pass the test. This is done by using the AND() function and then separating each individual test (of which there are two in this example, the vehicle type and the color).

- *Secondly* you have told DAX that the second test (on the color) can be any of several possibilities (two in this example). You did this using the OR() function, and separating each individual test by a comma.

Although I prefer to build complex functions from the inside out - that is by adding all the required parentheses first and adding what goes inside them second - this is not an obligation. You are free to build DAX formulas in any way that works.

Armed with this knowledge, you can now build extremely complex logical tests on your data. If you are an Excel or Access power user, then the learning curve should be quite short as the principles and functions are similar to those that you are using already. If you have come from the world of programming, then the concepts are probably familiar. If you are just starting out, then just be prepared to spend a little time practicing, and above all, analyze the question that you want DAX to answer *before* starting to write the statement.

Logical Operators

If you are writing fairly simple logical statements than DAX has an alternative to the AND(), OR and NOT() functions. These are explained in Table 6-3.

Table 6-3. *DAX Logical Operators*

Operator	Description	Example
&&	And	[Color] = "Red" && [VehicleType] = "Coupe"
||	Or	[Color] = "Red" || [Color] = "Blue"

At this point you may be wondering why DAX has two different - yet complementary - ways of building logical tests on data. Well, it largely depends on how many alternatives you want to handle in a logical evaluation. OR(), AND(), and NOT() functions can only accept *two* parameters. This is great if you are looking for red or blue cars but is less useful if you require red, blue, or silver cars. Or at least, using the OR(), AND(), and NOT() functions can get more complex if you need multiple alternatives.

Once again an example will probably make this easier to understand. Let's use both methods to flag red, blue, or silver cars with a comment using an IF() function. These will be new columns in the Vehicle table.

The first new column uses the following formula based on nesting OR() functions:

```
Best Selling Colors =
IF(OR([Color] = "Red", OR([Color] = "Blue", [Color] = "Silver")), "Best
Seller Color")
```

The second new column uses this piece of DAX

```
Best Selling Colors 2 =
IF([Color] = "Red" || [Color] = "Blue" || [Color] = "Silver", "Best
Seller Color")
```

You can see this working in Figure 6-8.

Figure 6-8. OR, AND, and NOT functions vs. operators

As you can see, the results are the same. So the approach is largely a question of personal preference.

The following apply to both approaches:

- You can mix logical tests across tables – however you will need to use the RELATED() function.

- You can apply AND, OR, and NOT conditions using elements from multiple fields. That is, you could create a test (also added to the Vehicle table) such as

```
Best Selling Combinations =
IF(OR([Color] = "Red", OR([VehicleType] = "Saloon", [Make]
= "Jaguar")), "Best Seller Combination")
```

Note   This example only showed three alternatives for the or operator. In practice you can add dozens of alternative conditions to these functions if you need to. What is important to remember is that you will have to *repeat the field* (and possibly the table name if it is not the current table) for *each comparison* – just like you did here when testing the colors of the cars.

The SWITCH() Function

The SWITCH() function is particularly useful when you want to

- Translate numbers that deliver a classification or categorization into a more comprehensible comment

- Translate TRUE or FALSE in a field to a more comprehensible comment

- Convert a code received from a source system into something more intuitive

- Make a cleaner looking equivalent to nested IF() statements

Standard SWITCH()

As an example of the use of SWITCH() to convert a number to a predefined comment, let's take the Stock Category Classification field that attributed a value between 1 and 4 to the number of weeks in stock, and presume that each of these values needs to appear in a separate column as a comment.

SWITCH() works like this:

- *First element* - This is the field that is to be used as the basis for applying an output.

- *Multiple pairs of elements* - This takes the value in the field used as the first parameter and delivers an output that varies according to the value that is input.

Here is how you can apply the SWITCH() function in practice:

1. In the PrestigeCarsDimensionalWithColumnsAcrossTables.pbix file, click on the *SalesInfo* table in the Fields list, and click the New Column button in the Table tools ribbon.

2. Enter the formula:

```
Stock Category Comment =
SWITCH([Stock Category Classification],
                1, "Something fishy going on!",
                2, "Good - only a short time in stock",
                3, "Acceptable period in stock",
                4, "Bad - Who ordered this car?"
    )
```

3. Press Enter or click the tick icon in the Formula bar. The new column will display the appropriate comment.

The output should look something like that shown in Figure 6-9.

Vehicle Age Category ▾	Vehicle Age Category Sort ▾	Stock Category Comment ▾
>30	7	Good - only a short time in stock
>30	7	Good - only a short time in stock
>30	7	Good - only a short time in stock
>30	7	Something fishy going on!
>30	7	Acceptable period in stock
>30	7	Acceptable period in stock
>30	7	Something fishy going on!
>30	7	Bad - Who ordered this car?

Figure 6-9. *SWITCH() output*

As you can see, SWITCH() is a lot easier to implement than nested IF() statements. The essential techniques that you need to remember when using SWITCH() are

- Enter a comma after the field to test.

- Enter commas between and after the matching value (1, 2, 3, or 4 in this example).

- Do *not* enter a comma after the final output element.

- You can add logical functions or logical operators inside a SWITCH() function.

You can add as many pairs of elements (lookup value and output) as you want to a SWITCH() function.

Note   A simple SWITCH() function can only be used for direct comparisons of values – that is where one value equals another. So you cannot use it to test if a value is greater than, less than, or not equal to another value. You will see a solution to this challenge in a couple of pages time.

One useful tweak that is worth knowing is that you can add a final output element (the comments in this example) to display a value in cases where the list of lookup elements does not map to a value in the lookup column. You can see this in the final line of the following piece of DAX (which needs to be added as a new column to the Geography table):

```
Geo Focus =
SWITCH([Town],
         "London", "UK Focus",
         "Birmingham", "UK Focus",
         "Paris", "France Focus",
         "N/A"
     )
```

You can see this working in Figure 6-10.

```
1 Geo Focus =
2 SWITCH([Town],
3         "London", "UK Focus",
4         "Birmingham", "UK Focus",
5         "Paris", "France Focus",
6         "N/A"
7     )
```

CountryName	GeographySK	Town	PostCode	SalesRegion	TownAbbrevation	Number of Sales Per Postcode	HasPostCode	Geo Focus
Spain	6	Madrid		EMEA	Mad	11	NoPostCode	N/A
France	7	Marseille		EMEA	Mar	32	NoPostCode	N/A
Italy	8	Milan		EMEA	Mil	11	NoPostCode	N/A
United States	9	New York		North America	New	6	NoPostCode	N/A
France	10	Paris		EMEA	Par	31	NoPostCode	France Focus
Germany	11	Stuttgart		EMEA	Stu	9	NoPostCode	N/A
United States	12	Washington		North America	Was	8	NoPostCode	N/A
Italy	13	Rome	120	EMEA	Rom	2	HasPostCode	N/A
Italy	14	Rome	129	EMEA	Rom	1	HasPostCode	N/A
Italy	15	Rome	175	EMEA	Rom	2	HasPostCode	N/A
Spain	16	Barcelona	8120	EMEA	Bar	7	HasPostCode	N/A
Spain	17	Barcelona	8400	EMEA	Bar	7	HasPostCode	N/A
Spain	18	Barcelona	8550	EMEA	Bar	2	HasPostCode	N/A
United Kingdom	19	Birmingham	B1 4BZ	EMEA	Bir	3	HasPostCode	UK Focus
United Kingdom	20	Birmingham	B1 7AZ	EMEA	Bir	8	HasPostCode	UK Focus
United Kingdom	21	Birmingham	B2 8UH	EMEA	Bir	6	HasPostCode	UK Focus

Figure 6-10. *Applying a "not found" alternative output to SWITCH()*

SWITCH() with TRUE()

Technically, SWITCH() can only test for exact matches between the values stored in the source column and those used in the SWITCH() function. There is, however, a tweak that can be used to test ranges of values. It is probably easier to see this in action and then explain it.

In the SalesInfo table, create a new column containing the following DAX code (you must have created the *Stock Category Classification* column described earlier in the chapter).

```
Stock Category Comment (range) =
SWITCH(TRUE(),
            [Stock Category Classification] < 2, "Bad",
            [Stock Category Classification] < 4, "OK",
            [Stock Category Classification] < 100, "Good"
      )
```

You should see the output shown in Figure 6-11.

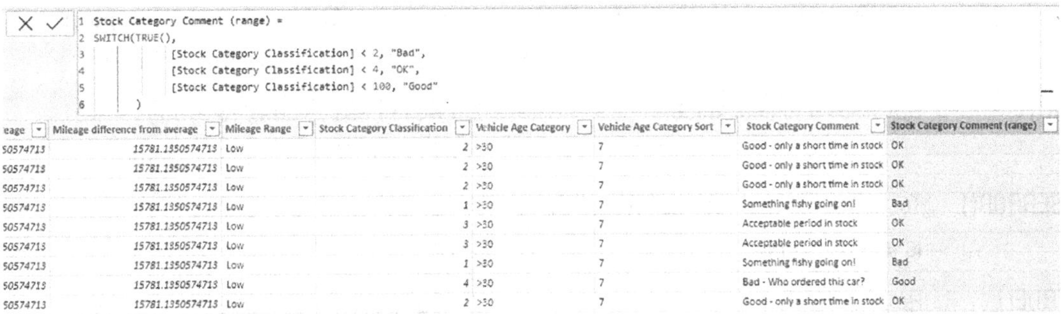

```
1 Stock Category Comment (range) =
2 SWITCH(TRUE(),
3             [Stock Category Classification] < 2, "Bad",
4             [Stock Category Classification] < 4, "OK",
5             [Stock Category Classification] < 100, "Good"
6       )
```

eage	Mileage difference from average	Mileage Range	Stock Category Classification	Vehicle Age Category	Vehicle Age Category Sort	Stock Category Comment	Stock Category Comment (range)
50574713	15781.1350574713	Low	2	>30	7	Good - only a short time in stock	OK
50574713	15781.1350574713	Low	2	>30	7	Good - only a short time in stock	OK
50574713	15781.1350574713	Low	2	>30	7	Good - only a short time in stock	OK
50574713	15781.1350574713	Low	1	>30	7	Something fishy going on!	Bad
50574713	15781.1350574713	Low	3	>30	7	Acceptable period in stock	OK
50574713	15781.1350574713	Low	3	>30	7	Acceptable period in stock	OK
50574713	15781.1350574713	Low	1	>30	7	Something fishy going on!	Bad
50574713	15781.1350574713	Low	4	>30	7	Bad - Who ordered this car?	Good
50574713	15781.1350574713	Low	2	>30	7	Good - only a short time in stock	OK

Figure 6-11. *SWITCH() with TRUE()*

When using SWITCH() with TRUE() (just like when using nested IF() statements), it is *vital* that you ensure that the comparison (the sequence of tests on the numeric value in the column that contains the source value to convert to a text) is *sequential.* So you *must* enter all the comparison values in ascending order. This is because SWITCH() returns the first value that matches a comparison. So, in this case, it applies "Bad" for any value less than 1, and then moves on to apply "OK" to any value less than 3 - unless already handled by the "less than 1" comparison.

DAX Logical and Information Functions

So far in this chapter you have seen a few of the DAX logical functions. In practice you may well need to build formulas that are built using some of the other functions that DAX provides to apply logic and to test the state and type of information in columns. Table 6-4 describes the essential functions that you may find yourself using as you create more complex data models.

Table 6-4. *DAX Logical Functions*

Operator	Description	Example
IF()	Tests a condition then applies a result if the test is true, and possibly a result if the test is false	IF([PartsCost]> 500, "Check Parts", "OK")
AND()	Extends the logic to include several conditions *all* of which must be met	IF(AND([PartsCost]> 500,[LaborCost]>1000), "Repair Cost Excessive", "OK")
OR()	Extends the logic to include several conditions *any* of which must be met	IF(OR([PartsCost]> 500,[LaborCost]>1000), "Repair or Labor Cost issue", "OK")
NOT()	Reverses a logical value	IF(NOT([PartsCost]> 500), "No Repair or Labor Cost issue", "")
ISERROR()	Tests a value and returns TRUE if there is an error value	IF(ISERROR([PartsCost]), "Check parts", "")
TRUE()	Returns TRUE	IF(SalesInfo[Mileage] > 100000, TRUE(), FALSE())
FALSE()	Returns FALSE	IF(SalesInfo[Mileage] > 100000, TRUE(), FALSE())
ISNUMBER()	Detects if a column value is numeric	IF(ISNUMBER([PartsCost]), "", "Data Error")
ISTEXT()	Detects if a column value is a text	IF(ISTEXT([PartsCost]), "Data Error", "")
ISNONTEXT()	Detects if a column value is not a text or is blank	IF(ISNONTEXT([PartsCost]), "", "Data Error")
ISODD()	Detects if a value is an odd number	IF(ISODD([PartsCost]), "Data Error", "")
ISEVEN()	Detects if a value is an even number	IF(ISEVEN([PartsCost]), "", "Data Error")
ISLOGICAL()	Detects if a column value is a true or false	IF(ISLOGICAL([PartsCost]), "Data Error", "")

Formatting Logical Results

Sometimes a logical function might exist only to return a simple true or false. For instance, you could want to test a value and indicate if it is over a certain threshold, using a formula like the following (added to the SalesInfo table):

```
High Mileage =
IF(SalesInfo[Mileage] > 100000, TRUE(), FALSE())
```

This formula simply tests the mileage figure for each record and returns TRUE() if the mileage is greater than 100,000 miles and FALSE() in all other cases.

However you might not want to display simply True or False in the column. So DAX will also let you format logical output - whether it is calculated like it is here or imported as a true or false value from a data source. The formula that you just saw can be formatted like this:

```
High Mileage =
FORMAT(
    IF(SalesInfo[Mileage] > 100000, TRUE(), FALSE()),
      "Yes/No"
    )
```

There are only three logical formats available in DAX. These are explained in Table 6-5.

Table 6-5. *DAX Logical Operators*

Format Code	Description
Yes/No	Formats the output as either Yes or No
True/False	Formats the output as either True or False
On/Off	Formats the output as either On or Off

> **Note**   Different data sources represent True in different ways. However nearly all represent False as a zero. Consequently Dax will interpret a logical column as a number and interpret any zeros as a false, and anything else as a True. If required, you can convert TRUE/FALSE to 1/0 just by wrapping the column in the INT function.

Safe Division

I imagine that, if you are an Excel user, you have seen your fair share of DIV/0 (divide by zero) errors in spreadsheets. Fortunately the Power BI Desktop team share your antipathy to this particular issue, and they have endowed Power BI Desktop with a particularly elegant solution to the problem. This is the DIVIDE() function. This function requires multiple parameters. These are

- The *numerator* - that is the number that will be divided by another number

- The *denominator* - the number that will be used to divide the first value

- The *error value* - the number that will be used if a divide by zero error is encountered

To see this in action, suppose that you want to add a new column that divides the contents of a column by the contents of another. Still using the sample file PowerBiDataModel.pbix, you can implement safe division like this:

1. In the sample file
   PrestigeCarsDimensionalWithColumnsAcrossTables.pbix and
   with the *Sales* table selected, activate the Table tools ribbon, and
   click the New Column button.

2. Enter the formula

   ```
   SalePriceToSalesCostsRatio =
   DIVIDE([DirectCosts],[SalePrice], 0)
   ```

3. Confirm the formula.

4. In the Table tools ribbon, click the percent button, then add a couple of decimals using the decimals button. As this metric is a ratio, it is best presented as a percentage to be more easily comprehensible not only in the current table but also in any visualizations that it appears in.

Used in a simple table (alongside the Direct Costs and SalePrice fields), it gives the output that you can see in Figure 6-12. As this column returns a ratio, I have formatted it as a percentage.

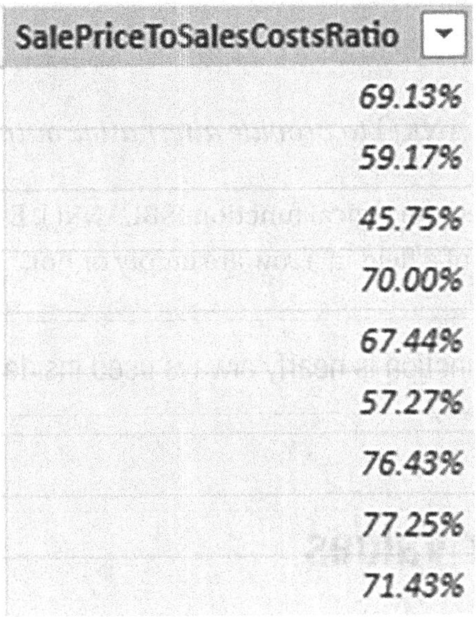

SalePriceToSalesCostsRatio
69.13%
59.17%
45.75%
70.00%
67.44%
57.27%
76.43%
77.25%
71.43%

Figure 6-12. *Safe division with DIVIDE()*

Testing for Blank or Empty Values

DAX can check if a column is blank for any row in the table. This can be useful when you want to avoid carrying out a calculation (or a part of a calculation), for instance.

As an example, try creating a new column in the SalesInfo table using the following DAX snippet:

```
New Buyer Comments =
IF(ISBLANK([BuyerComments]), "No Comment Provided", [BuyerComments])
```

What this formula does is to provide an alternative output for any blank elements in the BuyerComments column. The result is the output that you can see in Figure 6-13.

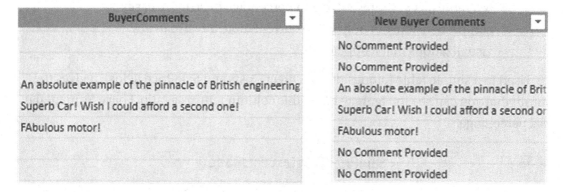

Figure 6-13. *Using ISBLANK() to provide alternative output*

This formula introduced the logical function ISBLANK(). Essentially, ISBLANK() tests whether the contents of a field in a row are empty or not.

Note   The ISBLANK() function is nearly always used inside an IF() function.

Testing for Error Values

You can check for calculations that return an error just as you can test for blank "cells" in a column with the ISERROR() function. Just like the ISBLANK() function, the ISERROR() function is nearly always used inside an IF() function.

Here, then is a simple piece of DAX that you can add to the Sales table as a new column

```
ErrorTrapping =
IF(ISERROR([CostPrice] / [TotalDiscount]), 0, [CostPrice] /
[TotalDiscount])
```

This particular formula is more for educational purposes than of any real use. However, if dividing one column by another gives an error, the formula handles the error gracefully, and displays a 0 as the output instead of displaying #Error.

Conclusion

This chapter introduced you to basic logic in calculated columns. You learned how to test the values in a column and return different outputs depending on what a column contained.

This covered handling value thresholds, and subsequently multiple successive thresholds using nested IF() statements. Finally you saw how to write logical tests using the SWITCH() function as well as how to test for blank values.

The DAX columns explained in this chapter are available for you to review in the file PrestigeCarsDimensionalWithColumnsAndLogic.pbix

The next (and final) part of this introduction to calculated columns is to see how to handle date and time columns. This is the subject of the next chapter.

CHAPTER 7

Date and Time Calculations in Columns

It is virtually inevitable that, at some point in your analytics career with Power BI, you will need to look at how data evolves over time. So it is equally inevitable that you will need at some point to add calculations based on dates.

Fortunately, date and time calculations are one area where DAX really shines. So we will see how date and time calculations can be applied to calculated columns in this chapter.

There are, essentially, three approaches to data (and time) calculations that we will be examining:

- Hb Extracting date and time elements from a source date, time, or datetime field

- Calculating timespans between two dates or times

- Projecting durations into the past or future from a given date

There is also a fourth aspect to date and time calculations called Time Intelligence. However, this merits an entire chapter to itself and is explained in Chapter 14.

To continue the progression of learning, this chapter will build on the data model that you extended with a few new calculated columns in the previous chapter. To save you having to build - or rebuild - this model, it is available as the Power BI Desktop file PrestigeCarsDimensionalWithColumnsAndLogic.pbix.

Once again, the source files used in this chapter are available for download from the Apress website as described in Appendix A.

© Adam Aspin 2023
A. Aspin, *Pro DAX and Data Modeling in Power BI*, https://doi.org/10.1007/978-1-4842-8995-2_7

Date Calculations

Before leaping into the details of datetime calculations (I will use the concept of datetime to cover date, time, and data where the date and time are calculated), there are a few basic considerations to take into account.

The first - and most important point - is that dates must be *recognizable as dates to DAX*. This means that the underlying data type of a column that you are using for a datetime calculation must be any of

- Date

- Datetime

- A text that can be interpreted as a date

Any column used for a time calculation must be any of the following data types:

- Time

- Datetime

- A text that can be interpreted as a time

It follows (building on what you learned in Chapter 2) that setting the appropriate data type in the data model will ensure that your datetime calculations stand a greater chance of being accurate. To begin this chapter, let's presume that the data model is accurate (meaning that dates and times are set to the correct data type) and see how to deal with dates that are entered as text a little later in the chapter.

Extracting Date Elements

As a first step in the path to datetime analysis, Power BI Desktop can help you to group and isolate date and time elements in your data. As time (by which we nearly always mean dates, times, or even dates and times) is often a continuous stream of dates in a dataset, it can be useful to separate out the years, months, weeks, and even days in a table so that you can create visuals that group and aggregate records into these more comprehensible "buckets" of time elements. You can then display and compare data over years and months, for instance, in order to tease out the real insights that your raw data contains.

As a first example of how DAX can help you to categorize records using date-based criteria, let's imagine that you envisage creating a couple of charts. Firstly you need one that let you track sales over the years that Prestige Cars has traded. Then you want a second graphic that looks at sales for each month over the years. Unfortunately (for the moment at least) your data model does not contain columns that show the year or the month of a sale - and Power BI Desktop does not let you create metrics as part of a visualization. You have to have the metric available in the data model if you plan to use it in a dashboard element. Indeed, this is precisely the reason that you are learning how to extend the data model with new columns using DAX. So here is how to create these two new columns in the SalesInfo table.

1. Open the file
    PrestigeCarsDimensionalWithColumnsAndLogic.pbix.

2. Switch to Data view.

3. Select the *SalesInfo* table in the Fields pane.

4. In the Table tools ribbon, click the New Column button.

5. In the Formula Bar to the left of the equals sign, replace Column
    with Sale Year.

6. Click on the right of the equals sign.

7. Enter (or start typing and then select) **YEAR(**.

8. Enter a left square bracket - **[**.

9. Select the field SaleDate.

10. Add a final right parenthesis to complete the YEAR() function.

11. Confirm the formula by pressing Enter or clicking the tick icon
    in the Formula Bar. The new column will display the year of
    each sale.

12. Repeat steps 2 through 9, only use the DAX formula MONTH()
    instead of YEAR(), and name the new column **Sale Month**.

The DAX formulas for the two new columns will be

```
Sale Month = MONTH([SaleDate])
Sale Year = YEAR([SaleDate])
```

You can see In Figure 7-1 what the SalesInfo table looks like with these two new columns added.

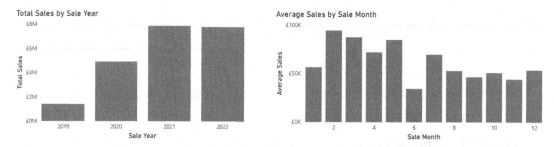

Figure 7-1. *The YEAR() and MONTH() DAX functions*

Either of these columns can now be used as the basis for charts that display sales over time as shown in Figure 7-2.

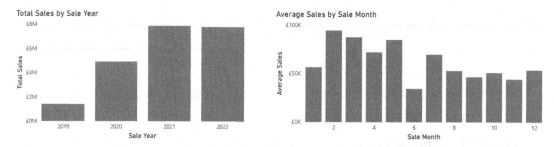

Figure 7-2. *The Year() and Month() DAX functions applied to a chart visual*

Of course this is only a beginning. Moreover, I would not suggest that you even should use these approaches in a complex data model - adding a Date dimension as described in Chapter 13 is more likely the best approach. Yet if you need a quick solution or if a date dimension is overkill, then using DAX data functions can deliver efficient results.

Note   To extract part of a date like this, the column that you are using for the original data must either be of the date data type, or be capable of being interpreted as a date by Power BI Desktop.

Date Elements

Once you have seen the core principles, you can extract any of the date elements from a date or datetime field that you can see in Table 7-1.

Table 7-1. *DAX Date Functions*

Function	Description	Example
YEAR()	Extracts the year element from a date	YEAR([SaleDate])
QUARTER()	Returns the calendar quarter of the date	QUARTER([SaleDate])
MONTH()	Extracts the month number from a date	MONTH([SaleDate])
DAY()	Extracts the day number from a date	DAY([SaleDate])
WEEKDAY()	Extracts the weekday from a date. Sunday is 1, Monday is 2, etc.	WEEKDAY([SaleDate])
WEEKNUM()	Extracts the number of the week in the year from a date	WEEKNUM([SaleDate])
EOMONTH()	Returns the last day of the month from a date after a specified number of months	EOMONTH([SaleDate], 2)
TODAY()	Returns the current date	TODAY()
TIME()	Lets you enter a time as hours minutes and seconds	TIME(19, 57, 25)
DATEDIFF()	Calculates the difference between two dates and/ or times expressed as a number of specified periods	DATEDIFF([SaleDate], DATE(2025, 07, 25), YEAR)
EDATE()	Returns the date a specified number of months after the given date (or before for a negative second parameter)	EDATE([SaleDate], 6)
UTCTODAY	Returns the current UTC date	UTCTODAY()
UTCNOW	Returns the current UTC date and time	UTCNOW()
YEARFRAC	Returns the year fraction of whole days between two dates	YEARFRAC([DateBought], [SaleDate], 3)
DATE()	Allows you to enter a date as year, month, and day	DATE(2022, 7, 25)

You can, of course, combine several DAX date (or time) functions into a single calculated column if you wish. The following formula (also added to the SalesInfo table) displays the year and month with a separator:

```
Sale Month and Year =
YEAR([SaleDate]) & " - " & MONTH([SaleDate])
```

I am not going to give detailed samples of all these functions here, as Chapter 13 explains how to use many of them to create a Date dimension. So you will find extensive examples of how to use date extraction DAX a little later in the book.

Extracting Time Elements

Extracting time elements from a datetime or time field is very similar to extracting date elements (or date parts as they are often called).

As the approach is so similar, I will not detail every step, but instead show you a complete DAX time function that displays the hour a sale took place. You can try this by creating a new column in the SalesInfo table that uses the following DAX:

```
Sale Time = HOUR([SaleDate])
```

You can see In Figure 7-3 part of the SalesInfo table with this column added.

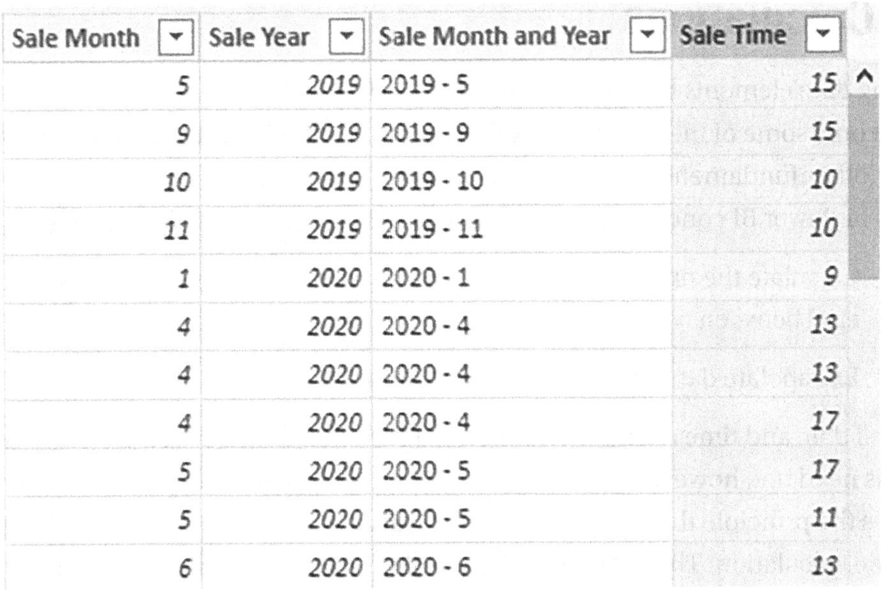

Sale Month ▼	Sale Year ▼	Sale Month and Year ▼	Sale Time ▼
5	2019	2019 - 5	15
9	2019	2019 - 9	15
10	2019	2019 - 10	10
11	2019	2019 - 11	10
1	2020	2020 - 1	9
4	2020	2020 - 4	13
4	2020	2020 - 4	13
4	2020	2020 - 4	17
5	2020	2020 - 5	17
5	2020	2020 - 5	11
6	2020	2020 - 6	13

Figure 7-3. *The HOUR() DAX functions*

Armed with these principles, you can now extract any of the time elements from a time or datetime field that you can see in Table 7-2.

Table 7-2. *DAX Time Functions*

Function	Description	Example
HOUR()	Extracts the hour from a time or datetime column	HOUR([SaleDate])
MINUTE()	Extracts the minutes from a time or datetime column	MINUTE([SaleDate])
SECOND()	Extracts the seconds from a time or datetime column	SECOND([SaleDate])
NOW()	Returns the current date and time	NOW()
TIME()	Lets you enter a time as hours minutes and seconds	TIME(19, 57, 30)
TIMEVALUE	Lets you enter a time as hours and minutes as a text	TIMEVALUE("20:19:57")

Date Calculations

Extracting date elements from source columns and converting certain source data to dates are only some of the functionality that you can obtain through applying DAX to dates. Another fundamental (and extremely useful) aspect of calculated date and time columns in Power BI concerns date calculations. This is because DAX enables you to

- Calculate the number of elapsed time periods (years, months, days, etc.) between two dates

- Extrapolate dates into the past or the future

Not all date and time elements enter the Power BI data model as the correct data type. This need not, however, prevent you from applying datetime calculations. Power BI applies the principle that if it can interpret a value as a datetime, then it will apply a datetime calculation. This is because Power BI (or rather DAX) will try and interpret any column as a datetime data type if you use a date or time function. It is worth noting that DAX will use the localization settings of the data model (which normally means the locale of your laptop) as the basis for interpreting dates. So, for example, if the locale is set to "US," then the date 3/8/2022 will be interpreted as the 8th of March 2022. If the locale is "GB," then the same date will be interpreted as the 3rd of August 2022.

DAX will always attempt to convert a text to a date if you apply a date function to a source column.

Note   Some text columns containing dates will, inevitably, contain data that can be interpreted by Power BI Desktop as a date or time – and other rows that cannot.

Setting Dates in a Calculation

Before looking at date calculations, it is worth reviewing the DAX functions that let you set a date inside a calculation. This enables you to calculate dates relative to either a fixed date that you enter or to the current date. There are four very simple functions that you need to understand:

- The NOW() function

- The TODAY() function

- The DATE() function

- The DATEVALUE() function

The NOW() and TODAY() Functions

The NOW() and TODAY() Functions give the current date. This enables DAX formulas to return a date calculation relative to the current date. The difference between the two is that

- The TODAY() function gives the current date

- The NOW() function gives the current date and time

Setting a Specific Date

In some date calculations you will need to enter a date as a point of reference for a date calculation. The DATE() function allows you to enter the separate parts of a date – either as a number or as a reference to a field or part of a field.

To see this in action, add two new columns to the SalesInfo table and apply the following two formulas:

```
Start Date = DATE(2018, 12, 31)
Start Date (Text) = DATEVALUE("31/12/2018")
```

You will see in Figure 7-4 that two columns are created containing a fixed date - and that both functions give the same result. This presumes that the formulas were created on a laptop where the locale is set to Europe.

Start Date ▼	Start Date (Text) ▼
31/12/2018 00:00:00	31/12/2018 00:00:00
31/12/2018 00:00:00	31/12/2018 00:00:00
31/12/2018 00:00:00	31/12/2018 00:00:00
31/12/2018 00:00:00	31/12/2018 00:00:00
31/12/2018 00:00:00	31/12/2018 00:00:00
31/12/2018 00:00:00	31/12/2018 00:00:00
31/12/2018 00:00:00	31/12/2018 00:00:00
31/12/2018 00:00:00	31/12/2018 00:00:00
31/12/2018 00:00:00	31/12/2018 00:00:00
31/12/2018 00:00:00	31/12/2018 00:00:00

Figure 7-4. *The DATE() amd DATEVALUE() DAX functions*

As you can see in Figure 7-4, these columns are only simplistic examples of how to use the DATE() and DATEVALUE()functions and might not be totally practical, but you will see in a few pages how to use them in more realistic contexts. What is important is that both are *valid datetime data types* ready for date and time calculations.

The DATE() Function

The DATE() function requires you to enter three parameters *in a strict order:*

- *First* the year
- *Second* the month number
- *Finally* the day number

You can enter the day and month numbers with leading zeros if you want. You can even add double quotes to enclose each separate element - but this is superfluous.

As an alternative to the DATE() function you can write, if you prefer:

```
Start Date = dt"2018-12-31"
```

The DATEVALUE( ) Function

The DATEVALUE () function only requires a single parameter. This is a text that can be interpreted as a date. The key points to remember when using this function are

- The date that you enter *must* be enclosed in double quotes

- The data that you enter *must be capable of being interpreted as a date* by DAX

The second point needs a little further explanation. DAX is very forgiving when it comes to converting a text to a date. Providing that DAX can recognize the date format that you have entered as a valid date, then the column will be converted to a datetime data type to be used in date calculations. So even dates expressed as texts such as

- "3-12-2020"

- "3 12 2020"

- "12/2020"

can be interpreted as dates. However these dates will be interpreted as MM-DD-YY (or MM DD YY) *whatever the locale* of Power BI Desktop.

The final takeaway is that you are always best entering a date as you would write it normally in the region that you work in. So in the United States you would write dates as MM/DD/YYYY and in Europe as DD/MM/YYYY. In any case it is probably advisable to test any non-standard format to see how DAX will interpret it. Alternatively you can avoid the problem altogether and write the date in ISO8601 format (YYYY-MM-DD).

Assembling Usable Dates

Sometimes, however, a date may have been imported from a source system in a format that cannot be interpreted as a valid date. The YYYYMMDD format is an example of something that DAX cannot interpret "out of the box," so you need to extend the DAX to recreate a valid date from the source data like this:

1.   With the *Sales* table selected, activate the Table tools ribbon, and click the New Column button.

2. Enter the following DAX formula:

```
New Date =
DATEVALUE(
        RIGHT([DateSK],2) & "/"
        & MID([DateSK],5,2) & "/"
        & LEFT([DateSK],4)
)
```

Confirm the formula. You can see the output for the DATE() and DATEVALUE() functions in Figure 7-5.

Figure 7-5. *The DATE() and DATEVALUE() functions*

There are three parts to this formula:

- *Firstly* – you used the functions LEFT(), MID(), and RIGHT() that you discovered in Chapter 4 to extract the parts of this text that correspond to year, month, and day.

- *Then* – you added date separators between these elements – which in effect made the source appear in the DD/MM/YYYY format.

- *Finally* – you wrapped the formula inside the DATEVALUE() function. This converts a text into a data data type.

Adding or Subtracting Dates

As befits a powerful formula language that is designed to aid in business analysis, DAX can calculate the time between two dates expressed as

- Years

- Months

- Weeks

- Days

- Quarters

- Hours

- Minutes

- Seconds

This can be extremely useful when you want to classify records according to a duration, and can be calculated using the DATEDIFF() function. The DATEDIFF() function expects you to apply *three* parameters when calculating an interval. These are

- The start date for the calculation of the interval.

- The end date up until when the interval will be calculated.

- The interval to calculate. This could be in years, days, or minutes, for example.

As an example, imagine that you want to display the number of days that each vehicle remained in stock, as this will help you to determine the fastest-selling models and consequently optimize the company's cash flow. As the data contains both the date bought for each vehicle as well as its sale date (even if the two are in different tables), this can be done using the DAX DATEDIFF() function.

1. In the PrestigeCarsDimensionalWithColumnsAndLogic.pbix file, click on the Data View icon, and then click on the *SalesInfo* table in the Fields list.

2. Enter the following formula (shown, along with the new column in Figure 7-5):

```
Days in Stock = DATEDIFF([DateBought],[SaleDate],DAY)
```

3. Press Enter or click the tick box in the formula bar. The formula will be added to the entire new column as you can see in Figure 7-6.

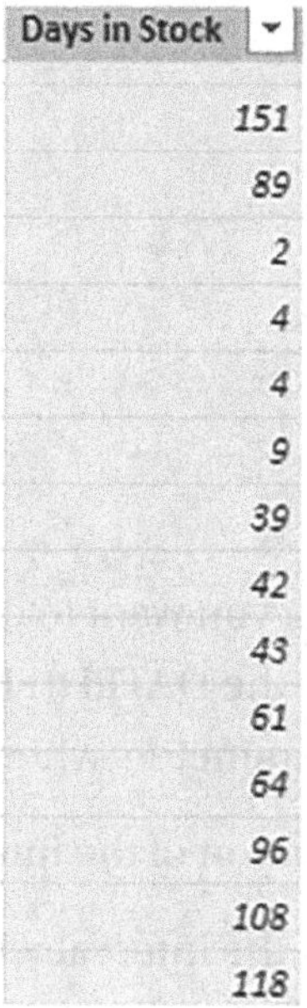

Days in Stock ▾
151
89
2
4
4
9
39
42
43
61
64
96
108
118

Figure 7-6. *The DATEDIFF() function*

You can now see the number of days that each vehicle was in stock before being sold and use this figure in Power BI Desktop dashboards. Only *complete* intervals will be displayed. In other words if you have selected YEAR as the interval, then the difference between the two dates must be in separate years for the function to return 1.

As you saw in the popup in step 10, the DATEDIFF() function lets you choose from a range of available intervals. These are explained in Table 7-3.

Table 7-3. *Date Difference Intervals*

Function	Description
YEAR	Returns the time difference in complete years
QUARTER	Returns the time difference in complete quarters
MONTH	Returns the time difference in complete months
WEEK	Returns the time difference in complete weeks
DAY	Returns the time difference in complete days
HOUR	Returns the time difference in complete hours
MINUTE	Returns the time difference in complete minutes
SECOND	Returns the time difference in complete seconds

Expressing the Difference Between Two Dates as Year and Month

Durations can be expressed in combinations of the various date elements such as years, months, and days. This involves extracting different date elements and then recombining them.

The following code snippet displays the difference between the purchase and sale dates as months and years. You need to add this as a new column to the SalesInfo table.

```
Years and Months To Sale = DATEDIFF([DateBought],[SaleDate],YEAR) & " Yrs "
& MOD(DATEDIFF([DateBought],[SaleDate],MONTH), 12) & " Mths"
```

You can see the result that this delivers in Figure 7-7.

0 Yrs 0 Mths
0 Yrs 0 Mths
0 Yrs 0 Mths
0 Yrs 1 Mths
0 Yrs 7 Mths
0 Yrs 0 Mths
1 Yrs 1 Mths
0 Yrs 11 Mths
0 Yrs 0 Mths
0 Yrs 1 Mths
0 Yrs 7 Mths
0 Yrs 3 Mths
0 Yrs 3 Mths

Figure 7-7. *Extending the DATEDIFF() function*

What is interesting here is the use of the MOD() operator. This returns the *remainder* of a division. Here we are dividing the number of months by 12 and returning the remaining number of months. So the formula shows the complete years and any month difference between the two dates.

Note   This approach is somewhat simplistic as it merely detects that month boundaries have been crossed. So buying a car on the 30th of November and selling it on the 1st of December will count as a month.

Of course, formulas like this one can be extended to present the results with greater accuracy and precision. However this is not the place in your DAX learning curve to add such levels of complexity.

Extrapolating Dates

Just as you can calculate elapsed date and time elements between two dates, you can project date and time intervals into the past and future.

As an example of this, imagine that PrestigeCars has a golden rule that no vehicle should be in stock for more than 90 days. So you want to add a column that gives the date 90 days after the purchase date in order to analyze stock turnover.

The DAX formula that you need to add to a new column in the SalesInfo table is

```
90 Day Limit = [SaleDate] + 90
```

The new column will automatically be a datetime data type.

This simple calculation makes the point that dates are, under the covers, numbers – where each number is one day. So you can add or subtract days by entering positive or negative numbers. In other words, If you are subtracting days, it is as simple as adding a minus sign before the figure for the number of days.

You can see the result of this formula in Figure 7-8. This column could, for instance, be used as part of a logic test to add a column that indicates cars that have spent too long on the forecourt.

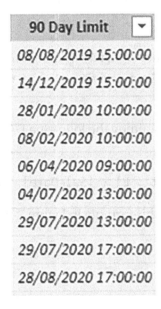

Figure 7-8. The extrapolating dates in DAX

Note   You can set the data type of the 90 Day Limit Date column as a date to avoid displaying the time element.

Adding or Subtracting Months

Extrapolating months into the past or future is made a little more complex due to the fact that there are not always the same number of days in the month. DAX can get round this challenge fairly easily by using the EDATE() function. This function projects a certain number of months in time – but if the requested date is past the last day of the projected month, then the last day of the projected month is returned.

The EDATE() function requires two parameters:

- The date that you will be adding months to - or subtracting them from.

- The number of months to add or subtract

To see this in action

1. In the PrestigeCarsDimensionalWithColumnsAndLogic.pbix file, click on the Data View icon, and then click on the *SalesInfo* table in the Fields list.

2. Add a new column containing the following DAX formula.

```
Stock Turnover =
IF(EDATE([DateBought], 6) < [SaleDate], "OK", "Slow seller")
```

You can see the output from this formula in Figure 7-9.

Figure 7-9. *The DAX EDATE() function*

Note   If you are subtracting months, Simply add a minus sign before the second parameter.

Calculating Years

To continue with elementary DAX date formulas — and as an admittedly extremely simple example — I will presume that we need to calculate how long it is since the car was purchased relative to the current date. As the source data contains the date the vehicle was bought for each vehicle, this will not be difficult. So, what you have to do is

1. In the file PrestigeCarsDimensionalWithColumnsAndLogic.pbix, and with the SalesInfo table selected, activate the Table tools ribbon, and click the New Column button.

2. Enter the following formula

    ```
    YearsSincePurchase = INT((TODAY()-[DateBought]) /365)
    ```

3. Press Enter or click the tick box in the formula bar. The formula
    will be added to the entire new column as shown in Figure 7-10.

YearsSincePurchase ▼
3
3
3
3
2
2
2
2
2

Figure 7-10. *Calculating years in DAX*

The INT() function is wrapped around the calculation of the number of days for two
reasons:

- It converts the underlying date data type that any date calculation
  returns to a *number*.

- It removes any decimal part for the division to display entire units
  (years in this example).

Adding Time to a Datetime

It is really easy in DAX to add hours, minutes, or seconds to a date or datetime field so
that you can project a date or time forwards or backwards. This means using the TIME()
DAX function, which takes three parameters:

- The number of hours to add

- The number of minutes to add

- The number of seconds to add

You can see this in action as follows:

1. In the file PrestigeCarsDimensionalWithColumnsAndLogic.pbix,
    and with the SalesInfo table selected, activate the Table tools
    ribbon, and click the New Column button.

2. Enter and confirm the following DAX formula

    `Date Bought 2 hours later = [DateBought] + TIME(0,2,0)`

Note   The new column will automatically be set to the DateTime data type.

One point that you may find interesting is that you can enter any positive number
for the hours, minutes, and seconds. So, for instance, you can enter TIME(0,0,240) to
add 4 minutes (240 seconds). So you do not need to convert seconds to hours, minutes
and seconds to add time to a field. Indeed, you can enter apparently incoherent time
durations such as

`TIME(2,77,240)`

This will result in adding 3 hours and 21 minutes to the start date. That is 4 minutes
as seconds 1 hour and 17 minutes as minutes plus 2 hours.

Note   Going over 24 hours will not, however, adjust the data and will hit an upper
limit of 23 hours being added.

Date and Time Formatting

When dealing with dates and times in Power BI Desktop, you may not need to go to the
lengths of extracting a part of a date field, but may simply need to display a date in a
different way. Rather like Excel, DAX can help you to do this quickly and easily.

Suppose that you want to have the SaleDate field displayed in a specific date format
for use in certain visualizations. Here is how you can do this:

1. In the file PrestigeCarsDimensionalWithColumnsAndLogic.pbix,
    and with the SalesInfo table selected, make sure the Table tools
    ribbon is active, and click the New Column button.

2. In the Formula Bar enter the following code:

    ```
    UKDate = FORMAT([SaleDate], "D-MMM-YYYY")
    ```

3. Confirm the formula by pressing Enter or clicking the tick icon in
    the Formula Bar. The new column will display the sale date in a
    different format.

If you take a look at Figure 7-11, you will see what the newly formatted invoice data
field looks like in the new column.

Figure 7-11. *Applying the Format() function*

Note   To reformat a date like this, the column that you are using for the original
data must either be of the date data type, or be capable of being interpreted as a
date by Power BI Desktop.

The date format that you applied in this example was not predefined by DAX in any
way. In fact it was assembled from a set of available day, month, and year codes that you
can combine to create the date format that you want. Table 7-4 explains the codes that
are available.

Table 7-4. *Custom Date Formats*

Format Code	Description	Example
d	The day of the month	"d MMM yyyy" produces 2 Jan 2016
dd	The day of the month with a leading zero when necessary	"dd MMM yyyy" produces 02 Jan 2016
ddd	The three letter abbreviation for the day of the week	"ddd d MMM yyyy" produces Sat 2 Jan 2016
dddd	The day of the week in full	"ddd dd MMM yyyy" produces Saturday 02 Jan 2016
M	The number of the month	"dd M yyyy" produces 02 1 2016
MM	The number of the month with a leading zero when necessary	"dd MM yyyy" produces 02 01 2016
MMM	The three letter abbreviation for the month	"dd MMM yyyy" produces 02 Jan 2016
MMMM	The full month	"dd MMMM yyyy" produces 02 January 2016
yy	The year as two digits	"d MMM yy" produces 2 Jan 16
yyyy	The full year	"MMMM yyyy" produces January 2016
HH	The hours since midnight with leading zeroes	"HH:MM:SS" produces 12:02:04
MM	The minutes of the hour with leading zeroes	"HH:MM:SS" produces 12:02:04
SS	The seconds elapsed with leading zeroes	"HH:MM:SS" produces 12:02:04
H	The hours since midnight without leading zeroes	"H:M:S" produces 12:2:4
M	The minutes of the hour without leading zeroes	"H:M:S" produces 12:2:4
S	The seconds elapsed without leading zeroes	"H:M:S" produces 12:2:4

If you do not want to build your own date formats, then you can choose from the six predefined date and time formats that Power BI desktop has available. These are explained in Table 7-5.

Table 7-5. *Predefined Date and Time Formats*

Format Code	Description	Example	Comments
Short Date	The short date as defined in the PC's settings	`Format([SaleDate], "Short Date")`	Formats the date using figures only
Long Date	The long date as defined in the PC's settings	`Format([SaleDate], "Long Date")`	Formats the date with the month as a text – and the day of the week
Long time	The long time as defined in the PC's settings	`Format([SaleDate], "Long Time")`	Formats the datetime or time column with the hour and minutes of the day
Short time	The short time as defined in the PC's settings	`Format([SaleDate], "Short Time")`	Formats the datetime or time column with the hour and minutes and seconds of the day
Medium time	Displays a time in 12 hour format	`Format([SaleDate], "Medium Time")`	Formats the date in 12 hour format
General date	The long date and time as defined in the PC's settings	`Format([SaleDate], "General Date")`	Formats the date using figures and the hour and minutes and seconds of the day

It is worth noting that dates that are, so to speak, "hard formatted" in this way can be calculated – but any date extrapolations will be displayed as numbers (the number of days since 31st of December 1899). You can, however, change the data type of numbered outputs like these to data and datetime data types to make them appear as dates.

Conclusion

This chapter provided a short overview of the core date and time functions that are available in DAX.

You learned how to enter dates and extract data elements from both date data columns and text columns that contain dates. You then saw that you can analyze the difference between two dates by years, months and days, as well as extrapolating dates and using dates in calculations.

If you want to see these functions as part of a data model, you can take a look at the file PrestigeCarsDimensionalWithDateTimeColumns.pbix that is available with the downloadable source data for this book.

This concludes the introduction to calculated columns in DAX. Now it is time to push DAX to the next level and learn all about measures. This will form the essence of the remainder of this book.

CHAPTER 8

Introduction to Measures

Now that you have a thorough grounding in calculated columns, it is time to boost your DAX abilities by getting to grips with measures in DAX. Measures are a second - and arguably far more powerful - way of applying calculations in Power BI. They are, however, very different in their scope and application to column-based calculations. I cannot deny that they can be more difficult to understand and apply in many cases. Yet mastering measures is key to unleashing the full potential of Power BI as an analytics tool.

I will begin with the simpler aspects of writing measures and progress over the next few chapters to a deeper understanding of the capabilities that measures can add to your dashboards.

You will almost always need to add measures to your tables when you need to deliver more advanced and pliable calculations (or measures or metrics — call them what you will). Measures can produce some extremely powerful results to help you analyze your data. This is because measures do things that calculated columns simply cannot do. So if you need to work more with aggregate values and less on a row-by-row basis, then you will have to create measures to achieve the desired result.

Measures are different from calculated columns in the following ways:

- They are not linked to any specific table.

- They can be placed in any table.

- They can use fields from anywhere in the data model.

- They work across the entire data model (initially, anyway) using existing relationships and filter direction.

- They work over columns (largely) not rows.

- They aggregate and filter data.

© Adam Aspin 2023
A. Aspin, *Pro DAX and Data Modeling in Power BI*, https://doi.org/10.1007/978-1-4842-8995-2_8

- They modify and override the interactions of filters and slicers on the result.

- They react to changes in filters and slicers to produce dynamically calculated results.

While adding new columns can provide much of the extra data that you want to output in tools like Power BI Desktop, it is unlikely that this approach can deliver *all* the analyses that you need. Specifically, calculated columns can *only* work on a row-by-row basis; they cannot contain formulas that have to apply to all or part of the records in a column. For instance, counting the number of cars sold for a year, a quarter, or a month has nothing to do with the data in a single row in the Sales table. It does, however, concern the table as a whole.

One fundamental point to bear in mind is that measures often need a lot of thought on how they will be used and what the ramifications of creating a calculation using a measure will be. This aspect of measure development will become more evident as you start to learn about measures and their use.

As with so many aspects of Power BI Desktop and self-service business intelligence in general, measures are probably best introduced through a few examples. It will, unfortunately, be impossible to do anything other than scratch the surface of measures in a single chapter as they are arguably the most powerful element in Power BI Desktop. Nonetheless, I hope that the short introduction in this chapter will whet your appetite, and that you will then want to continue to learn more about measures in the chapters that follow.

In this chapter we will continue to develop the Power BI Desktop file PrestigeCarsDimensionalWithDateTimeColumns.pbix file that you began in the previous chapter.

A First Measure: Number of Cars Sold

Suppose that you want to be able to display the number of cars ever sold by Prestige Cars. Not only that, but you want this figure to adjust when it is filtered or sliced by another criterion such as country or color. Put simply, you want this metric to be infinitely sensitive to how its display is influenced by filters and slicers - or which column and row it appears in - yet always give the right answer.

So how are we going to achieve this? Here is how:

1. Open the sample file
    PrestigeCarsDimensionalWithDateTimeColumns.Pbix.

2. Either in Data view or dashboard view, select the Table to which
    you wish to add a measure. I am choosing Vehicle in this example.

3. In the Modeling ribbon, click the New Measure button. The
    Formula Bar will look like Figure 8-1, and the Measure Tools
    ribbon will be displayed.

Figure 8-1. *The Formula Bar when creating a measure*

4. Add the following formula to the formula bar:

    ```
    NumberOfVehicleVariants =COUNTROWS(
    ```

5. The moment that you add the left parenthesis, you will see that
    the popup will then suggest a list of DAX formulas interspersed
    with the names of tables in the current data model. If you scroll
    down the list, it will look like Figure 8-2.

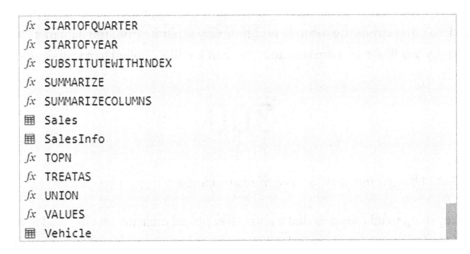

Figure 8-2. *The popup menu showing functions and tables*

6. Select Vehicle and add the right parenthesis. The formula will look like this:

```
NumberOfVehicleVariants = COUNTROWS(Vehicle)
```

7. Confirm the creation of the formula by pressing Enter or by clicking the check mark icon in the formula bar. A new field will appear in the Fields pane in the Vehicle table.

After having just learned about calculated columns, the first thing that will strike you initially is that *no column is created for this measure*. The only indication that it exists is its presence in the Fields List (once you expand the Vehicle table). If you look closely at the field (NumberOfVehicleVariants) that you have just created, you will also see that there is a tiny icon of a calculator to the left of the field name. You can see this in Figure 8-3.

Figure 8-3. *The measure icon in the Fields pane*

This allows you to distinguish measures from New Columns (which have a small Fx icon to the left of the field name).

Not difficult, I am sure you will agree. Yet the best is yet to come. Suppose that you now use this field in a Power BI Desktop, which is also filtered to show the results for 2019 only. The result will be filtered so that only the number of sales for 2019 is displayed. In other words the formula is completely separate from the data in a column, but will apply any filters that are selected. You can see this applied to a visualization in Figure 8-4.

300

NumberOfVehicleVariants

Figure 8-4. *Using a measure in a card visualization*

The key thing to take away is that a correctly applied measure can be used in a Power BI Desktop table, chart, or indeed any type of visualization and will always show the correct result of any and all filters and slicers that you have applied (assuming that all the required logic has been applied). Also the figures will be correct figure for each intersection of rows and columns in tables. So all in all, it is well worth ensuring that you have all the measures that you need for your analytical output in place and that they are working correctly in Power BI Desktop, because you can then rely on these calculations in the data set in so many different visualizations.

Tip   You can rename measures by either right-clicking on the name of the measure in the Fields list and selecting Rename, or by clicking on the name of the measure in the Fields list and altering the name in the Formula Bar. You can also do this by double clicking, or pressing F2.

When you start out creating measures, it can be a little disconcerting at first not to see the results of a formula immediately, as you can when adding new columns. If this worries you (or if you want to test the result of a new measure), then one approach is to create a table in Dashboard view and add the new measure plus any other useful measures that allow you to verify that everything works as you expected.

Note   You can add a measure to any table. You do not have to add a measure to a table where the fields you are using as a measure's parameters reside as measures can use fields from any table in the data model.

Basic Aggregations in Measures

Measures are DAX formulas, so in learning to use measures you will have to become familiar with some more DAX functions. My intention here, though, is definitely not to take you instantly through all that measures can offer. They are too complex and powerful for that. Instead I would like to show you a few basic formulas that can be useful in real-world dashboards, and give you some initial DAX recipes that should prove practical.

So, as a second example, let's calculate total costs of vehicles purchased. Although you can just type in a simple DAX formula, I prefer to show you how you can extend the knowledge that you gained when creating calculated columns and apply many of the same techniques to creating measures.

1. In Power BI Desktop (and still in the file PrestigeCarsDimensionalWithDateTimeColumns.pbix), ensure that you are in Data view.

2. Select the Vehicle table, and click the New Measure button in the Table Tools ribbon.

3. Replace Measure with the name that you want to use (Total Sales).

4. Click to the right of the equals sign (=).

5. Enter SUM as the function, followed by a left parenthesis. A list of all the tables and fields in the data model will appear (including any columns and measures that you have added).

6. Enter a left bracket to restrict the popup list to fields in the current table.

7. Start typing the field name (SalePrice in this example). After a couple of characters, any tables or fields with these characters will be listed, as shown in Figure 8-5.

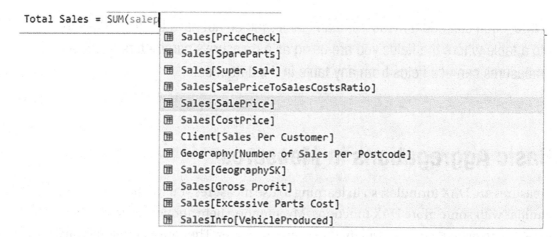

Figure 8-5. *Creating a measure containing an aggregation*

8. Scroll down and select the [SalePrice] field.

9. Add a right parenthesis. The formula should read:

```
Total Sales=SUM('Sales'[SalePrice])
```

10.   Press Enter (or click the check mark icon in the formula bar).

The measure will be created and appear in the fields list. This particular function will give you the total of the SalePrice column. However, when you use it in Power BI Desktop, it will be filtered and applied (or sliced and diced if you prefer) to take into account the subset of data currently in scope.

A final point is that when you insert a table or a field from the popup list that you can see in Figure 8-5, you will see three types of icon to the left of the table or field: the icon with a table outline denotes a Power BI Desktop table; the table with a selected column icon indicates a column of data or a column that you have added.

Note   In step 7, when selecting rather than entering a measure you can press the ' (single quote) key instead of the left square bracket to list the tables as well as the fields that you are looking for.

To practice a little try creating the average, maximum, and minimum sale price using the formulas in Table 8-1.

Table 8-1. *A Few Elementary DAX Measures*

Name	DAX Code
Average Cost Price	AVERAGE('Sales'[CostPrice])
Maximum Sale Price	MAX('Sales'[SalePrice])
Minimum Sale Price	MIN('Sales'[SalePrice])

Default Measures

Creating a measure to represent the sum of the sale price was, in some ways, superfluous. This is because the SUM() of any numerical value is created automatically, under the covers in Power BI Desktop, whenever you drag a numeric field into a visual. So you have been using aggregations in measures for as long as you have been using Power BI - only without realizing it.

This snippet of knowledge reinforces the point that, outside calculated columns, each time you drag a numeric field into a visual, you are aggregating it under the covers. In contrast, when you are creating measures, you have to be explicit about the aggregation that you are creating.

It is worth remembering that you can modify the aggregation that Power BI uses in any visual simply by clicking on the popup menu chevron for the field in the Visualizations pane. You can see this in Figure 8-6.

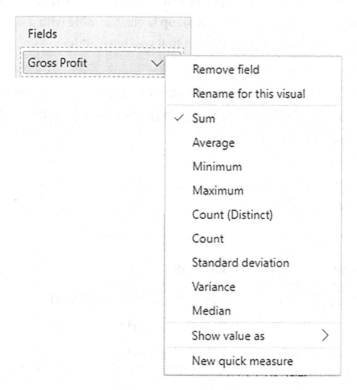

Figure 8-6. *Setting the aggregation to use in a visual*

If you choose a different aggregation, then Power BI simply applies the selected aggregation function behind the scenes.

Measures Are Column-Based Calculations

It is fundamental to understand that measures apply to entire columns or tables of data. While a calculated column may refer to a column (or field if you prefer), the field is taken to be specific for each row in the calculated column. This is emphatically *not* the case for a measure. Using a field reference in a measure refers to the entire column of figures, irrespective of the row.

So measures - at least at this stage in the learning curve - *must aggregate the data*. So you *have* to wrap any field referenced in a measure in an aggregation function (again, at least for the moment in your apprenticeship of measures in DAX). This allows DAX to apply the calculation that you are applying to the field. This is part of the power of measures, as they are independent of actual rows.

If you try and create a measure simply be referring to the column in the table, say by using a formula like

```
Total Without Aggregation = 'Sales'[CostPrice]
```

Then the measure will simply not work and will return an error message like the one shown in Figure 8-7.

A single value for column 'CostPrice' in table 'Sales' cannot be determined. This can happen when a measure formula refers to a column that contains many values without specifying an aggregation such as min, max, count, or sum to get a single result.

Figure 8-7. *Creating a measure without an aggregation*

While the error message is not, at least in my opinion, totally clear, it is trying to tell you that *DAX measures cannot refer simply to a field*. As DAX does not know which row to use in a measure, you have to aggregate field data in some way. Much of what is to come in the chapters on measures will explain ways of doing exactly this.

Ways to Create Measures

As befits a feature that is so fundamental to data model and dashboard development, there are multiple ways of telling Power BI Desktop that you want to add a measure. These include

- The New Measure button in the Home ribbon in the Report or Data Views

- The New Measure button in the Table Tools ribbon in the Report or Data Views

- The New Measure button in the Modeling ribbon in the Report View

- The New Measure button in the Measure Tools ribbon (which is only visible if you have selected an existing measure in the Fields pane)

- Right-clicking a table in the Fields pane name and selecting New measure in the context menu

- Clicking the ellipses to the right of a table name in the Fields pane and selecting New measure in the context menu

Measures can be added in both Report view and Data view. They *cannot* be added in Model view.

Modifying Measures

Measures are modified in exactly the same way as calculated columns.

1. Click on the measure name in the Fields pane. The measure contents will appear in the Formula bar.

2. Edit the measure in the Formula bar.

3. Confirm your modifications.

Field References

A couple of pages back we created a simple measure named TotalSales. This formula could also have been written as

```
TotalSales=SUM([SalePrice])
```

This formula works - but it is not really "best practice" when it comes to using field references inside DAX measures. This is because a field reference should, ideally, include the table name. This is purely to assist you, the DAX developer, when you come back to revise a complex measure at a later date. So in this case the formula should read

```
TotalSales = SUM('Sales'[SalePrice])
```

What is more, if a table name contains spaces, then the table name needs to be in single quotes. In all cases the field name has to be enclosed in square brackets.

Composite Measures

As you can well imagine, not all metrics that you will be using are likely to be as simple as those that you just saw. You can, of course, create measures that are the result of combining several DAX functions (and multiple references to different fields) to achieve the result that you are looking for.

For a little practice, then, you could try adding the ratio of gross profit to sale price to the data model. This measure can then be used to create dashboards with Power BI Desktop. What is more, this upcoming measure also makes the point that you can refer to data columns or calculated columns in measures in exactly the same way.

1. In Power BI Desktop (still using the sample file PrestigeCarsDimensionalWithDateTimeColumns.pbix), ensure that you are in Data view.

2. Select the *Sales* table and click the New Measure button in the Table Tools ribbon.

3. Replace Measure with the name that you want to use (RatioGrossProfit).

4. Click to the right of the equals sign (=).

5. Enter SUM as the function, followed by a left parenthesis. You will see all the available tables and fields in the data model.

6. Select the field Sales[Gross Profit].

7. Enter a right parenthesis.

8. Enter a forward slash (the divide by operator).

9. Enter SUM as the function, followed by a left parenthesis.

10. Enter a left bracket to limit the list to fields in the current table.

11. Select the field Sales[SalePrice].

12. Enter a right parenthesis. The formula should read

    ```
    RatioGrossProfit = SUM('Sales'[Gross Profit])/
    SUM('Sales'[SalePrice])
    ```

13. Press Enter (or click the check mark icon in the formula bar).

14. In the Measure Tools ribbon, click the percentage button to apply a percentage format.

Note   You could, of course, use the DIVIDE() function to perform the division if you prefer. Indeed, using DIVIDE() is usually considered best practice. In this case the formula would read

---

```
RatioGrossProfit =
DIVIDE(SUM([Gross Profit]), SUM([SalePrice]))
```

The easiest way to see the output of a measure like this is to create a table that uses the measure and another attribute so that you can see how the measure works in practice. Figure 8-8 shows a simple example of this - also for 2020.

Color	SalePrice	CostPrice	Gross Profit	Ratio Gross Profit
Black	£1,251,410.00	£1,021,150.00	£230,260.00	18.40%
Blue	£423,650.00	£362,644.00	£61,006.00	14.40%
British Racing Green	£663,250.00	£440,398.00	£222,852.00	33.60%
Canary Yellow	£287,070.00	£176,824.00	£110,246.00	38.40%
Dark Purple	£283,000.00	£226,400.00	£56,600.00	20.00%
Green	£370,750.00	£263,974.00	£106,776.00	28.80%
Night Blue	£405,500.00	£295,204.00	£110,296.00	27.20%
Pink	£235,950.00	£173,282.00	£62,668.00	26.56%
Red	£442,500.00	£326,318.00	£116,182.00	26.26%
Silver	£542,205.00	£410,773.00	£131,432.00	24.24%
Total	**£4,905,285.00**	**£3,696,967.00**	**£1,208,318.00**	**24.63%**

Figure 8-8. *Applying a measure in a table*

This is, of course, an extremely simple example of a composite DAX function. Indeed, you can probably see a distinct resemblance to an Excel formula. However the key point to take away is that once you have created the measure, it will work in just about any Power BI visualization and using most, if not all, of the attributes from the data model. Power BI can also apply the measure intelligently to hierarchies of data. So, for instance, if you add the Make field to the preceding table and switch the visualization to a matrix, you will instantly see the calculations that are shown in Figure 8-9. Here, Power BI Desktop has calculated automatically the Gross Profit Ratio for each vehicle sold without you having to alter the formula in any way.

Make	SalePrice	CostPrice	Gross Profit	Ratio Gross Profit
⊟ Alfa Romeo	£51,050.00	£36,637.00	£14,413.00	28.23%
Black	£12,500.00	£10,200.00	£2,300.00	18.40%
British Racing Green	£21,050.00	£13,977.00	£7,073.00	33.60%
Green	£17,500.00	£12,460.00	£5,040.00	28.80%
⊞ Aston Martin	£1,191,470.00	£877,728.00	£313,742.00	26.33%
⊞ Bentley	£732,140.00	£554,106.00	£178,034.00	24.32%
⊞ Ferrari	£1,083,450.00	£836,472.00	£246,978.00	22.80%
⊞ Jaguar	£299,915.00	£214,424.00	£85,491.00	28.51%
⊞ Lamborghini	£542,150.00	£422,871.00	£119,279.00	22.00%
⊞ Porsche	£582,600.00	£431,126.00	£151,474.00	26.00%
⊞ Rolls Royce	£279,600.00	£208,799.00	£70,801.00	25.32%
⊞ Triumph	£142,910.00	£114,804.00	£28,106.00	19.67%
Total	£4,905,285.00	£3,696,967.00	£1,208,318.00	24.63%

Figure 8-9. *A hierarchy using a measure*

When you use measures like this one in visualizations, you may well find that some calculations are displayed to many decimal places. If you find this distracting, then you can format measures in the same way that you format Power BI Desktop columns. Any formats that you apply will be used in Power BI Desktop by default whenever you use this measure.

Cross-Table Measures

You are not limited to creating measures that refer to the fields in a single table. If anything, measures are designed to apply across *all* the fields in a data model. To start out with a simple example, suppose that you want to create a custom measure that displays the ratio of the selling price to the vehicle mileage (apparently the sales director considers this a key metric).

1. In Power BI Desktop ensure that you are in Report or Data view (the top two of the view icons on the left below the ribbon). This example is still using the sample file PrestigeCarsDimensionalWithDateTimeColumns.pbix.

2. Select the *SalesInfo* table, and click the New Measure button in the Modeling ribbon.

3. Enter the following formula:

```
SalePriceToMileage =
SUM(Sales[SalePrice]) / SUM(SalesInfo[Mileage])
```

4. Press Enter (or click the check mark icon in the formula bar).

You could then add this new measure to a matrix. If you do, you will see something like Figure 8-10.

Make	SalePrice	CostPrice	Gross Profit	SalePriceToMileage
⊟ **Alfa Romeo**	**£243,510.00**	**£194,808.00**	**£48,702.00**	**25.4%**
Black	£93,125.00	£74,500.00	£18,625.00	24.0%
Blue	£46,450.00	£37,160.00	£9,290.00	17.2%
British Racing Green	£21,050.00	£16,840.00	£4,210.00	21.1%
Dark Purple	£21,500.00	£17,200.00	£4,300.00	64.7%
Green	£17,500.00	£14,000.00	£3,500.00	52.6%
Night Blue	£35,190.00	£28,152.00	£7,038.00	35.3%
Red	£8,695.00	£6,956.00	£1,739.00	26.2%
⊟ **Aston Martin**	**£5,102,755.00**	**£4,082,204.00**	**£1,020,551.00**	**131.3%**
Black	£1,439,560.00	£1,151,648.00	£287,912.00	137.6%
Blue	£355,500.00	£284,400.00	£71,100.00	116.5%
British Racing Green	£259,385.00	£207,508.00	£51,877.00	95.4%
Canary Yellow	£266,350.00	£213,080.00	£53,270.00	126.7%
Dark Purple	£114,600.00	£91,680.00	£22,920.00	114.9%
Green	£599,590.00	£479,672.00	£119,918.00	235.8%
Night Blue	£662,090.00	£529,672.00	£132,418.00	112.2%
Pink	£263,100.00	£210,480.00	£52,620.00	263.8%
Red	£402,900.00	£322,320.00	£80,580.00	138.8%
Silver	£739,680.00	£591,744.00	£147,936.00	102.7%
⊟ **Austin**	**£64,500.00**	**£51,600.00**	**£12,900.00**	**19.1%**
Black	£29,650.00	£23,720.00	£5,930.00	20.3%
Canary Yellow	£2,250.00	£1,800.00	£450.00	2.4%
Night Blue	£23,600.00	£18,880.00	£4,720.00	71.0%
Total	**£21,977,950.00**	**£17,574,360.00**	**£4,403,590.00**	**128.8%**

Figure 8-10. *Applying multiple measures to a matrix*

Creating cross-table measures in DAX is, if anything, easier than creating new columns that use values from more than one table. As you can see from this example,

- You do not need to use the RELATED() or the RELATEDTABLE() functions.

- You only have to specify the table name before entering the field name (or select the combination of table and field from the popup).

- You can use the Fields List to refer to fields in other tables (or even in the same table).

- You *must* use aggregation functions on numeric fields. If you do not, you will get an error message.

- When editing a measure, you can make the list of available *fields and measures* appear by pressing the ' (single quote) character.

- When editing a measure, you can make the list of available *measures* appear by pressing the [(left square bracket) character.

- You can narrow the list of available fields and measures by entering a few characters that are contained anywhere in the required field once the popup list of available fields is displayed.

Note   A measure is attached to a table so that it will appear as a field in the specific table. The measure does not have to use any of the fields in the table that "hosts" it. This means that you can attach measures to any table in your data model which allows for a considerable organizational freedom when extending the model with further metrics. Indeed, adding measures to a table (or tables) that contains only measures is often considered best practice. This approach was explained in Chapter 3.

Cascading Measures

In Chapter 4 you saw that you can use a calculated column in a DAX formula that creates another calculated column - and that this creates a "Domino effect" where modifications to a formula in an "upstream" calculated column flow down to any downstream DAX formulas that refer to the original column. This is a key aspect of DAX, as it allows you to define formulas that can be re-used by other formulas. This avoids you having to copy and paste (or rewrite) complex DAX code that becomes both hard to maintain and a source of potential errors.

The same approach applies to measures. So you can create measures that refer to previously created measures. Let's see an example of this.

What I want to do here is to rework the RatioGrossProfit measure so that the SalePrice and GrossProfit are themselves measures. This means that these measures will already be aggregated when they are used in the RatioGrossProfit measure.

Let's see how this works in practice - even if it means creating a couple of additional simple measures:

- TotalSalePrice

- TotalGrossProfit

1. In Power BI Desktop (still using the sample file PrestigeCarsDimensionalWithDateTimeColumns.pbix), ensure that you are in Data view.

2. Select the Sales table, and click the New Measure button in the Table Tools ribbon.

3. Enter the measure

   TotalSalePrice = SUM(Sales[SalePrice])

4. Select the Sales table, and click the New Measure button in the Table Tools ribbon.

5. Enter the measure

   TotalGrossProfit = SUM(Sales[Gross Profit])

6. Still in the Sales table, click the New Measure button in the Table Tools ribbon.

7. Enter **RatioGrossProfit2** as the measure name.

8. On the right of the equals sign, enter (or start typing then select) DIVIDE(.

9. After the left parenthesis, press the [(left square bracket) key. The list of available measures will be displayed, as shown in Figure 8-11.

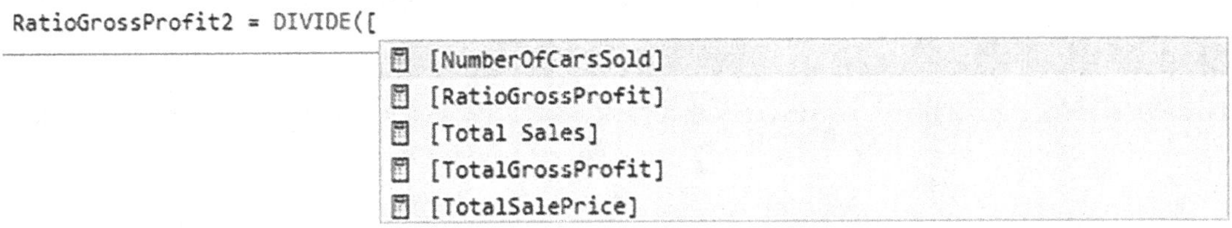

```
RatioGrossProfit2 = DIVIDE([
```
🗊 [NumberOfCarsSold]
🗊 [RatioGrossProfit]
🗊 [Total Sales]
🗊 [TotalGrossProfit]
🗊 [TotalSalePrice]

Figure 8-11. *Reusing existing measures*

10. Select the existing TotalGrossProfit measure.

11. Enter a comma.

12. Press the [(left square bracket) key to display all existing measures.

13. Select the existing TotalSalePrice measure.

14. Add a right parenthesis to finish the DIVIDE() function.

15. Press Enter (or click the check mark icon in the formula bar). Measure tools ribbon.

The formula should read

```
RatioGrossProfit2 = DIVIDE([TotalGrossProfit], [TotalSalePrice])
```

This formula will deliver exactly the same output as the previous RatioGrossProfit formula. However, if ever you need to modify the calculation for the total gross profit or the total sales, then you can carry out any required modifications once only, in the upstream formulas (TotalGrossProfit and Total Sales in this example). The change will then cascade downstream into every formula that refers to these "source" formulas.

If, as is the case with most Power BI users, you come from the Excel universe, then it is probably extremely reassuring to know that the principle of cascading calculations is a fundamental part of the DAX approach too.

So the principle behind a comprehensible and elegant data model as far as DAX measures are concerned is

- Define core metrics so that they can be *reused* instead of rewriting them inside multiple different formulas.

- Format the core metrics.

- Use meaningful names to define measures. This will assist your users and help you when it comes to debugging and updating measures.

Implicit Filters Applied To Measures

Whenever you have added numeric fields to Power BI Desktop visuals, these fields have not only been implicitly aggregated (as I mentioned earlier) but have also been implicitly *filtered*.

As you begin your journey into developing DAX measures, it is vital to understand some elementary elements behind how DAX calculations work. While explaining all the subtleties will underpin the remainder of this book, as a starting point you need to remember that

- Each numeric element (a figure in a card, a "cell" in a table or a point in a chart) is calculated independently of all other calculations

- Each calculation is affected (not to say controlled) by all the factors that apply implicit filters. These include

  1. Dashboard filters (at report page and visual level)

  2. Slicers (unless the interaction has defined that a slicer does not affect the visual containing the calculation)

  3. Row or axis specification (the level of granularity of a row or axis and what defines this)

  4. Column or axis specification(the level of granularity of a column or axis and what defines this)

  To make this clearer, take a look at Figure 8-12 that shows how a calculation (and this applies to every calculation in the matrix) depends on the implicit filters that are applied.

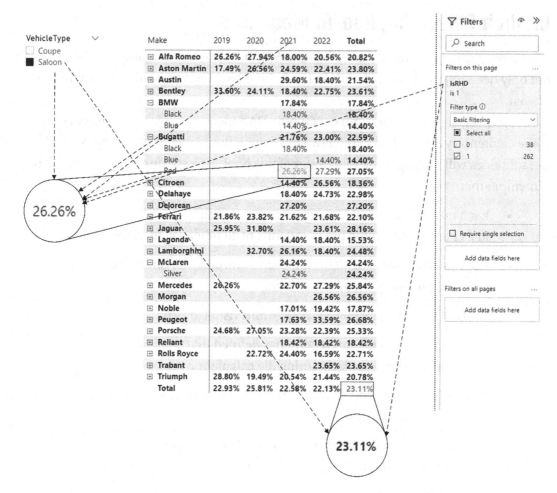

Figure 8-12. *Implicit filtering*

Here you can see that the highlighted calculation is filtered by five factors:

- The filter in the Filter Pane (is the car right hand drive)

- The slicer (the type of vehicle)

- The make (Bugatti)

- The color (red)

- The year (2021)

Totals by make are filtered by the Filter Pane, the slicer, and the year (except for the final column of totals where the year is not used to filter data).

Totals by year are filtered by the Filter Pane, the slicer, and the make (and model depending on the level of granularity) except for the final row of totals where the make and model are not used to filter data.

The grand total, however, is *only* filtered by the Filter Pane and the slicer.

As you move on with DAX, you will discover that much of what you will be doing is to replace the implicit with the explicit as far as filtering is concerned. That is you will be

- Specifying the actual calculation to apply instead of relying on a default aggregation

- Specifying the filtering that should be applied instead of letting implicit filtering apply

Naming Convention for Measures

Measures - like calculated columns - can have virtually any name you want. However, you are likely to go on to develop data models that contain dozens or even hundreds of measures. Consequently, a little logic in the names that you give them can make your life easier going forward.

Some elementary ideas that you may want to implement are

- *Core measures* – those that feed into other measures but are not necessarily going to be used as stand-alone calculations destined to be applied by other users (or appear as column titles); can have names that do not contain spaces or non-alphabetical characters (such as parentheses and punctuation).

- *User-focused measures* – those that are destined to be applied by end-users; should be immediately comprehensible and can contain spaces and non-alphabetical characters.

Annotate Measures

Measures can become extremely complex. Not only that, but it can be difficult when you return to a data model that you created months previously to remember all the subtleties of the logic or use of a specific measure.

So my advice is to explain measures by adding a description in the Properties pane that you can display in Model view. This need only takes a few seconds (which could be nothing in comparison with the time spent actually developing a measure) but can save you an immense amount of frustration later. Not only that, but a clear description can be harnessed by users of your data model to create their own dashboards faster - and without harassing you for explanations.

To add a description to a measure

1. Using the sample file
   PrestigeCarsDimensionalWithDateTimeColumns.pbix), switch to
   Model view.

2. In the Fields pane, click on the measure that you wish to annotate
   (Cost Plus Spares Margin in this example).

3. Expand the Properties pane if closed.

4. Add descriptive text in the Description area. You can see an
   example of this in Figure 8-13.

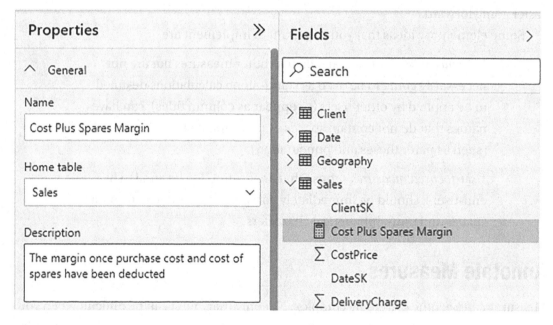

Figure 8-13. *Adding a description to a measure*

Now, when you or your users click on or hover over a measure in the fields list, the description will appear in a popup. You can see this in Figure 8-14.

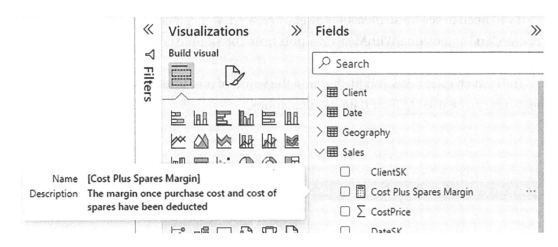

Figure 8-14. *Adding a description to a measure*

Measure Recalculation

Another area where measures differ from calculated columns is recalculation. Measures are recalculated on the fly whenever a filter or slicer is modified (or even when levels in a matrix are expanded or a measure is added to a visual). So there is absolutely no need to refresh the data for measures to refresh.

Conclusion

In this chapter you took a first look at one of the most powerful features in DAX - measures. These let you develop a custom calculations that you add to the Power BI Desktop data model. You then use these metrics in your visuals to deliver specific insights based on your data.

You saw that measures apply to entire columns or tables of data - and that this nearly always involves using aggregation functions in measures. You also learned that measures apply across the entire dataset. Finally you saw that measures can become the source for further measures, allowing you to create cascading calculations in a data model.

This chapter can only be a brief introduction to measures in DAX. Yet I hope that it has whetted your appetite and that you will now feel encouraged to continue with the learning curve so that you can move on to develop the analyses and metrics that you need using more advanced measures.

If you need to see these measures already created, you can open the file PrestigeCarsDimensionalWithMeasures.pbix from the sample data that is available to download.

The next chapter takes you further into the world of measures by explaining how measures can be used to filter data in calculations.

CHAPTER 9

Filtering Measures

One of the most powerful aspects of measures in DAX is that they can be tweaked and tuned to filter the output that they deliver. This means that you can use them to create targeted calculations that only display the results based on a *specific subset* of the data in the data model - rather than *all* the data defined by any implicit filter selection. This opens the doors to endless ways to compare and contrast results and consequently deliver in-depth analysis of your data.

The art behind the extension of measures is to apply *explicit* filters to measures. The aim is to write a measure that delivers a specific result based on isolating a subset of the data in the way that you have chosen. There are several approaches to applying explicit filters to measures in DAX. However, this chapter will begin with what is probably the core filtering approach - the CALCULATE() function.

CALCULATE() is, quite simply, the most powerful (and possibly complex) function in DAX. You saw it very briefly in Chapter 5 applied to calculated columns, but this chapter will introduce the CALCULATE() function in greater depth. We will not be looking at all that CALCULATE() can do in this chapter, however, as I prefer to explain the different aspects of this function gradually over several chapters. It is so powerful and complex that I feel that a more gradual series of explanations will be more conducive to appreciating the sheer range of possibilities that this function can bring to DAX. A more structured approach should also prevent newcomers to DAX becoming overwhelmed by the sheer complexity of the CALCULATE() function.

In this chapter we will continue to develop the Power BI Desktop file PrestigeCarsDimensionalWithMeasures.pbix file that you began in the previous chapter.

© Adam Aspin 2023
A. Aspin, *Pro DAX and Data Modeling in Power BI*, https://doi.org/10.1007/978-1-4842-8995-2_9

Filtering Data in Measures

There will, inevitably, be times when the filters that you apply using the Power BI Desktop user interface (the query-level filters that were described above) are not quite what you are looking for. There could be several reasons for this, including

- You want to apply a highly specific filter to a single metric

- You want to override the natural result of the query-level filter

- You are creating a highly complex formula and it has to be tailored to a specific use

Any of these reasons (and there are many others that you will discover as you progress with DAX) could require you to filter the data in a measure. Let's take a look at a few examples of circumstances where this could prove necessary.

Note   In this chapter we will be looking at what are often described as simple filters inside CALCULATE( ). These filters can deliver a wide range of practical answers to real-world analytics challenges. However, they are not the whole story. So be prepared to continue with the rest of the book if you do not find solutions to all your calculation challenges in this chapter.

Simple Filters

There will probably be many occasions in your career analyzing data when you need to home in on a specific subset of data. Maybe you need to compare and contrast one sales stream with another. Perhaps you need to highlight one cost compared to a total. Whatever the actual requirement, you will need to apply a specific filter to a metric.

There are dozens - if not hundreds - of ways of applying different filters when building measures. One of the basic approaches is to use the CALCULATE() function to apply a range of filters to a measure that you can then use in the visualizations that you build into your dashboards.

Text Filters

To begin, let's look at a fairly simple filter requirement. Prestige Cars sells to clients in many countries. As part of your ongoing sales analysis, you want to isolate the sales for the UK sales stream and compare sales for this destination to all sales. Here is how you can do this.

1. Click on the Sales table in the Fields List.

2. In the Modeling ribbon, click on the New Measure button.

3. In the Formula Bar replace Measure with **UK Sales**.

4. To the right of the equals sign, enter (or select) **CALCULATE(**.

5. Enter (or select) **SUM(**.

6. Select the field Sales[SalePrice].

7. Add a right parenthesis. This will terminate the SUM() function.

8. Enter a comma. This tells the CALCULATE() function that you are about to add the filters.

9. Enter a single quote to display all the available fields in the dataset.

10. Select the field 'Geography'[CountryName].

11. Enter an equals sign.

12. Add the words **"United Kingdom"** - including the double quotes before and after the country name.

13. Add a right parenthesis. This will terminate the CALCULATE() function. The formula should now read

```
UK Sales =
CALCULATE(SUM(Sales[SalePrice]), 'Geography'[CountryName] =
"United Kingdom")
```

You could then add this new measure to a simple table of sales by Make. If you do you, we will see something like Figure 9-1 (a slicer sets the year to 2020 only).

217

Make	SalePrice	UK Sales
Alfa Romeo	£243,510.00	£111,245
Aston Martin	£5,102,755.00	£2,653,965
Austin	£64,500.00	£32,400
Bentley	£1,689,240.00	£689,000
BMW	£60,500.00	£5,500
Bugatti	£1,915,500.00	£985,000
Citroen	£101,080.00	£2,340
Delahaye	£107,000.00	£39,500
Delorean	£99,500.00	£99,500
Ferrari	£5,540,850.00	£2,154,900
Jaguar	£882,955.00	£347,740
Lagonda	£218,000.00	£218,000
Lamborghini	£1,727,150.00	£382,150
Total	**£21,977,950.00**	**£9,211,375**

Figure 9-1. *Using a simple text filter in the CALCULATE( ) function*

You can see from this table that the SalePrice column is not explicitly filtered in any way. It is nonetheless implicitly filtered by the Make and Year. However the UK Sales column always shows a smaller figure for the sales per make, as it is displaying only the explicitly filtered subset of data that you requested as well as the implicit filters on Make and Year. The new measure that you created can be applied to any visualization and can be filtered, sliced, and diced like any other data column, calculated column, or metric.

Now let me explain. As you saw when building the formula, we are using the DAX CALCULATE() function. This function does what its name implies; it *calculates* an aggregation. However, the calculation is often - as is the case here - a *filter* operation. This is because of the way in which the CALCULATE() function works. It is based on two parameters:

- ***The first parameter*** – defines the *function* to use (SUM() in this example), and the table and column that is aggregated; it could have been potentially a much more complex formula.

- ***The second parameter*** – is a *filter* to force DAX to show only a subset of the data. In this specific case it returns the sum of sales *only* when the client is in the UK.

The filter that is applied here comes from another column. Indeed it comes from another table altogether. When using the CALCULATE() function, you can use just about any column (either an original data column or a calculated column) as the source for a filter. The essential caveat here is that the data model must not only join the tables (or allow for a join path through other tables) but must also enable a filter path through the table relationships.

CALCULATE() can apply many separate filters. So the "second" parameter could in fact be a whole series of separate parameters each of which applies different filters. You will look at more advanced filter parameters further on in this chapter.

Note   When you are filtering on a text (such as the country name in this example), you *must* always enclose the text that you are searching for in double quotes.

Numeric Filters

You are not restricted to filtering on text-based data only in Power BI Desktop. You can also subset data by numeric values. As an example, suppose that you want to see totals for sales for lower-priced models so that you can target the higher end of the market. Here is how you could do this.

1. Click on the *Sales* table in the Fields List.

2. In the Modeling ribbon, click on the New Measure button.

    ```
    LowPriceSales =
    CALCULATE(SUM(Sales[SalePrice]), Sales[SalePrice] < 50000)
    ```

Comparing the low price sales with the unfiltered sales per make for 2020 extends the table you created previously to give a result like the one shown in Figure 9-2.

Make	SalePrice	UK Sales	LowPriceSales
Alfa Romeo	£243,510.00	£111,245	£243,510
Aston Martin	£5,102,755.00	£2,653,965	£1,042,715
Austin	£64,500.00	£32,400	£64,500
Bentley	£1,689,240.00	£689,000	
BMW	£60,500.00	£5,500	£60,500
Bugatti	£1,915,500.00	£985,000	
Citroen	£101,080.00	£2,340	£35,190
Delahaye	£107,000.00	£39,500	£107,000
Delorean	£99,500.00	£99,500	
Ferrari	£5,540,850.00	£2,154,900	
Jaguar	£882,955.00	£347,740	£489,955
Lagonda	£218,000.00	£218,000	
Lamborghini	£1,727,150.00	£382,150	£28,650
Total	**£21,977,950.00**	**£9,211,375**	**£3,792,840**

Figure 9-2. *Using a numeric filter inside the CALCULATE( ) function*

In this simple example, you saw how to use the Less Than (<) comparison operator. You can, of course, use any of the standard logical comparison operators (=, <. >, <=, >=, <>) that were explained in Chapter 6.

Note   When you are filtering on a number, you must *not* enclose the number that you are searching for in quotes. Neither must you format the number in any way.

Boolean (True/False) Filters

CALCULATE() can filter on Boolean (or true/false) values as well. This is dependent on the data types of the column that you are filtering being set to the true/false data type.

You can define Boolean filters like this:

- Client[IsCreditRisk] = TRUE()

- Client[IsCreditRisk] = FALSE()

So you could add the following measures to any table (in the sample files you can find these in the SalesInfo table):

```
Creditworthy Sales =
CALCULATE(SUM(Sales[SalePrice]),
Client[IsCreditRisk] = TRUE())
```

```
UnCreditworthy Sales =
CALCULATE(SUM(Sales[SalePrice]),
Client[IsCreditRisk] = FALSE())
```

For Boolean filters this can be simplified as follows:

```
Creditworthy Sales =
CALCULATE(SUM(Sales[SalePrice]), Client[IsCreditRisk])
```

```
UnCreditworthy Sales =
CALCULATE(SUM(Sales[SalePrice]), NOT Client[IsCreditRisk])
```

In this example, simply placing a true/false field as a parameter in a CALCULATE() measure is equivalent to writing Table[Field] = TRUE().

You can see these measures applied to a table, a matrix, and a chart in Figure 9-3. In the matrix I have renamed the two measures to save space.

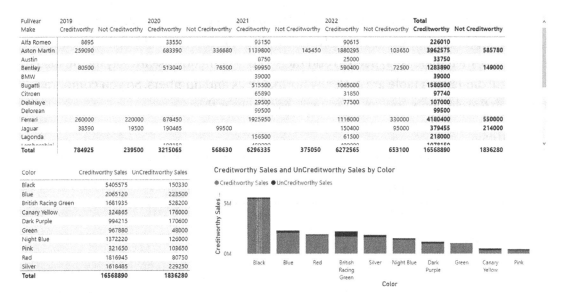

Figure 9-3. Boolean filters in a CALCULATE( ) function

As you can see from this simplistic dashboard, the measures can be used in all kinds of visuals. Moreover they accept further implicit filtering by row, column, and axis.

Filtering Dates

Dates, of course, can be filtered in measures using a CALCULATE() function. Inevitably, this implies that the column that you are filtering on is a Date or DateTime data type.

To see all the car sales for cars purchased after the 1st of January 2020, you could write DAX like this:

```
Sales 2020 And Beyond =
CALCULATE(SUM(Sales[SalePrice]),
'SalesInfo'[DateBought] >= DATE(2020, 1, 1))
```

You do not have to use the DATE() function to enter a date if you do not want to. Instead you can enter a date as a text - and convert it to a date that DAX can use in filter comparisons using DAX like this (as you saw in Chapter 7):

```
Sales 2020 And Beyond 2 =
CALCULATE(SUM(Sales[SalePrice]),
'SalesInfo'[DateBought] >= DATEVALUE("01/01/2020"))
```

If you want to find these formulas in the sample files look in the SalesInfo table.

If you are using a date dimension (and I really hope that you will be after reading Chapter 13), then there is no need to add the VALUE() function as virtually all the fields in the date dimension table are perfectly normal texts and numbers. So you compare them with the elements you are filtering on as simply as this:

```
2020 Sales=
CALCULATE(SUM(Sales[SalePrice]),'Date'[FullYear] = 2020)
```

You can see these formulas applied in Figure 9-4.

Color	SalePrice	Sales 2020 And Beyond	2020 Sales
Black	£5,555,905.00	£5,423,365	£821,320
Blue	£2,288,620.00	£2,004,020	£193,200
British Racing Green	£2,210,135.00	£2,098,635	£663,250
Canary Yellow	£500,865.00	£500,865	£287,070
Dark Purple	£1,164,815.00	£1,164,815	£283,000
Green	£1,015,880.00	£799,280	£233,650
Night Blue	£1,498,220.00	£1,440,230	£405,500
Pink	£425,300.00	£425,300	£179,000
Red	£1,897,695.00	£1,765,500	£261,150
Silver	£1,847,735.00	£1,758,735	£456,555
Total	£18,405,170.00	£17,380,745	£3,783,695

Figure 9-4. *Date filters in a CALCULATE() function*

Time Filters

To complete the basic overview of using CALCULATE() to filter data, you need to see how to apply time filters. The following DAX (added to the Sales table as it concerns sales data) can isolate sales between 5 and 6 PM:

```
Five oClock Sales =
CALCULATE(AVERAGE('Sales'[SalePrice]), HOUR('SalesInfo'[SaleDate]) = 17)
```

This measure - like so many of the DAX measures that you have looked at in this chapter – also shows that many DAX functions such as date, time, and text functions can be used in a range of circumstances to solve a multitude of potential challenges. Here you are reusing the HOUR() function that you first saw in Chapter 7.

Creating a measure based on this code and then using it in a table gives the output that you can see in Figure 9-5.

Town	Five oClock Sales
Barcelona	£29,500.00
Berlin	£36,525.00
Birmingham	£72,333.33
Brussels	£17,000.00
Geneva	£9,950.00
London	£55,193.00
Los Angeles	£30,500.00
Lyon	£44,600.00
Manchester	£56,000.00
Marseille	£176,112.50
Milan	£12,500.00
New York	£56,900.00
Newcastle	£31,250.00
Paris	£99,725.00
Total	**£60,481.39**

Figure 9-5. *Using a time filter inside the CALCULATE( ) function*

More Complex Filters

In the two previous examples, you saw the basics of creating filtered measures using either a text, a number, a Boolean value, a date, or a time to subset the data returned by the CALCULATE() function. In the real world of data analysis, filters can get a lot more complex than these. Indeed, allowing you to create specific and complex filters for metrics is one of the ways that DAX can help you to tease out real insight from your data. So here are a few examples of the ways that you can define more complex filtered metrics in your data models to introduce you to a more in-depth look at ways of combining and extending filters.

Multiple Criteria in Filters

The CALCULATE() function is not limited to a single filter. Far from it. You can add multiple filters as the second parameter each separated by a comma. All the parameters will combine to filter the result.

Schematically, a CALCULATE() function can be considered as is shown in Figure 9-6. Please note that this is far from a complete definition of all that CALCULATE() can do, but it is a necessary next step in the learning curve.

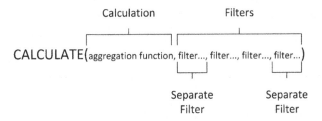

Figure 9-6. *The structure of CALCULATE( ) in complex filters*

Now to see this in practice. For example, imagine that you didn't just want to see UK sales when you look at your data, but also want to see the (slightly lower) figure for UK sales where the client has a good credit status. This means combining two filter criteria in a DAX snippet like this (in the Sales table):

```
Creditworthy UK Sales =
CALCULATE(SUM(Sales[SalePrice]),
'Geography'[CountryName] = "United Kingdom", Client[IsCreditRisk]=FALSE()
)
```

If this column is added to the table that you saw in Figure 9-5, you will see a result something like the one in Figure 9-7. Not surprisingly, only towns in the United Kingdom have creditworthy UK sales in this table.

Town	Five oClock Sales	Creditworthy UK Sales
Barcelona	£29,500.00	
Berlin	£36,525.00	
Birmingham	£72,333.33	£1,697,680
Brussels	£17,000.00	
Geneva	£9,950.00	
Liverpool		£1,225,125
London	£55,193.00	£1,606,865
Los Angeles	£30,500.00	
Lyon	£44,600.00	
Manchester	£56,000.00	£581,045
Marseille	£176,112.50	
Milan	£12,500.00	
New York	£56,900.00	
Newcastle	£31,250.00	£599,430
Paris	£99,725.00	
Stoke		£937,150
Total	**£60,481.39**	**£6,647,295**

Figure 9-7. *Using a more complex filter*

This was, if anything, a simple example. The filters that you add to any measure that uses the CALCULATE() function can contain multiple elements. Also you can create a series of filters where each filter element compares data from different tables and mixes both text-based, numeric, true/false, and date (or time or datetime) filters.

Moreover, if you cast your mind back to combining data from multiple tables in calculated columns in Chapter 5, you will have noticed that there is no need for the RELATED() or RELATEDTABLE() functions when defining filters (at this level at least) in measures. This confirms the point that measures can apply across the whole data model.

AND/OR Filters in Measures

Adding multiple parameters to a measure combines all the filters using *and* logic. That is, the restrictions applied by each filter produce a cumulative effect. However, inside an individual filter parameter, you can specify alternative filters using selection using OR filters. The approach taken is fairly similar to what you have already seen in Chapter 4 when filtering data in calculated columns, but it is nonetheless worth taking a quick look at how filters work in measures both to consolidate and to extend your knowledge.

Complementary Choices (AND Filters)

Logical AND filters are applied "out of the box" in measures where an explicit CALCULATE() is applied as each filter parameter is combined with all others using AND logic to narrow down the selection. So, to restrict an aggregation so that it only displays the total for cars of one color and make, you can write DAX like the following and add the measure to the Sales table:

```
Red Aston Martins Only =
CALCULATE(SUM(Sales[SalePrice]), 'Vehicle'[Color] = "Red", 'Vehicle'[Make] =
"Aston Martin")
```

Simply separating each complete logical filter using a comma is sufficient to apply multiple filters. Moreover, there is no limit to the number of filters that can be applied in this way. You can see the output from this measure in Figure 9-8.

Alternative Choices (OR Filters)

Each filter parameter can contain alternative options. For this you can use the same logical OR approaches that you discovered a few chapters ago (in Chapter 6 to be precise). As the approach is very similar, I will not provide an exhaustive look at all possible approaches here. Instead I will refresh your memory - and point out the similarities - using a few succinct examples.

Alternative Elements from a Single Column

For instance, to specify that the output should be restricted to either red or blue cars, you can use DAX like this:

```
Red or Blue Cars Sales =
CALCULATE(SUM(Sales[SalePrice]),
OR('Vehicle'[Color] = "Red", 'Vehicle'[Color] = "Blue"))
```

While the OR() *function* is fine for only two alternatives, the alternative format (the OR *operator*) is easier if there are more than two elements in a list of alternatives - as you can see in the following DAX snippet:

```
Red, Green or Blue Cars Sales = CALCULATE(SUM(Sales[SalePrice]),
'Vehicle'[Color] = "Red" || 'Vehicle'[Color] = "Green" || 'Vehicle'[Color] =
"Blue")
```

However the IN operator can be the easiest option for longer lists of alternatives as you only have to specify the field that you want to filter on once. You can see this in the following piece of DAX:

```
Specific Countries =
CALCULATE(SUM(Sales[SalePrice]),
'Geography'[CountryName] IN {"France", "United States", "Belgium",
"Switzerland"})
```

You can see these measures (and the one from the previous section) in action in Figure 9-8. I have added them to the Sales table for convenience.

Town	Red Aston Martins Only	Red or Blue Cars Sales	Red, Green or Blue Cars Sales	Specific Countries
Barcelona		104750	104750	
Berlin	£61,500	61500	61500	
Birmingham		1001450	1130100	
Brussels		19450	144450	311850
Geneva	£79,500	134950	143450	
Liverpool		99250	112200	
London		53925	261575	
Los Angeles		131150	375150	855690
Lyon		19950	75550	175650
Madrid		207340	219330	
Manchester		162300	162300	
Marseille	£45,500	480200	554650	3119225
Milan	£45,000	52750	52750	
New York				451740
Paris		1359350	1423850	2021235
Rome			82590	
Stoke		156500	156500	
Washington		141500	141500	422690
Total	£231,500	4186315	5202195	7358080

Figure 9-8. *CALCULATE( ) used for OR filters*

Note   When using the IN operator, be sure to place the list of elements inside curly braces. Each list element must be in double quotes, and the elements are comma-separated.

Alternatives Across Columns

If you want to widen the selection and use OR filters that give the choice between elements from different columns *in the same table*, you can use DAX like this:

```
Any Red OR Aston Martin =
CALCULATE(SUM('Sales'[SalePrice]),
'Vehicle'[Color] = "Red" || 'Vehicle'[Make] = "Aston Martin")
```

This filter will return the total sales whether the car is an Aston Martin or is red (or both).

Moreover, you can use lists of alternative elements (using the IN function followed by a comma-separated list inside curly braces) should you need to filter on multiple alternative elements.

You can add as many alternative filter elements from as many different columns (in the same table) as you want. So the previous code snippet can be extended like this:

```
Any Red OR Aston Martin OR Saloon = CALCULATE(SUM('Sales'[SalePrice]),
'Vehicle'[Color] = "Red" || 'Vehicle'[Make] = "Aston Martin" ||
'Vehicle'[VehicleType] = "Saloon")
```

You can see these measures applied in Figure 9-9.

CountryName	Any Red OR Aston Martin	Any Red OR Aston Martin OR Saloon
Belgium	£230,950	£287,400
France	£1,660,130	£5,125,195
Germany	£106,500	£425,440
Italy	£373,240	£946,330
Spain	£103,550	£1,492,510
Switzerland	£283,600	£559,765
United Kingdom	£2,700,060	£7,341,615
United States	£756,520	£1,516,970
Total	**£6,214,550**	**£17,695,225**

Figure 9-9. *CALCULATE( ) used in cross-column filters*

You can find these two measures in the Sales table of the sample files.

Excluding Elements (NOT Filters)

You can apply filter parameters in measures to define elements that must be excluded from the result as well. The following piece of DAX specifies "red or blue cars that are not Jaguars":

```
Red or Blue Cars Sales No Jaguars = CALCULATE(SUM(Sales[SalePrice]),
'Vehicle'[Color] = "Red" || 'Vehicle'[Color] = "Blue", 'Vehicle'[Make] <>
"Jaguar")
```

Pushing things further, you can request the total sales for "red or blue cars that are not Jaguars or Aston Martins":

```
Red or Blue Cars Sales No Jaguars or Aston Martins = CALCULATE(SUM(Sales[Sa
lePrice]),
'Vehicle'[Color] = "Red" || 'Vehicle'[Color] = "Blue", 'Vehicle'[Make] <>
"Jaguar", 'Vehicle'[Make] <> "Aston Martin")
```

As you can see here, adding multiple parameters to filter on the same field is perfectly permissible - though you could also write it like this if you prefer:

```
Red or Blue Cars Sales No Jaguars or Aston Martins 2 = CALCULATE(SUM(Sa
les[SalePrice]), 'Vehicle'[Color] = "Red" || 'Vehicle'[Color] = "Blue",
AND('Vehicle'[Make] <> "Jaguar", 'Vehicle'[Make] <> "Aston Martin"))
```

Finally, it is equally easy to define a filter that excludes a list of elements from the result. This means using the NOT ... IN syntax that you can see applied in the following DAX snippet:

```
Exclude Certain Countries =
CALCULATE(SUM(Sales[SalePrice]),
NOT 'Geography'[CountryName] IN {"France", "United States", "Belgium",
"Switzerland"})
```

The trick with this piece of code is to add the NOT operator before the field name that you are using in the filter (instead of placing it before the IN operator - which is what SQL programmers inevitably try and do).

You can see the result of applying these measures in Figure 9-10.

FullYear	Red or Blue Cars Sales No Jaguars	Red or Blue Cars Sales No Jaguars or Aston Martins	Red or Blue Cars Sales No Jaguars or Aston Martins 2	Exclude Certain Countries
2019	401195	364695	364695	767885
2020	424700	193200	193200	2464600
2021	1318920	1318920	1318920	3186760
2022	1909750	1590750	1590750	4068080
Total	4054565	3467565	3467565	10487325

Figure 9-10.   Using CALCULATE( ) to exclude elements

You can find these measures in the Sales table of the sample files.

Partial Text Filtering

Sometimes you may find yourself faced with the challenge of filtering not on the entire field but on part of the text in a field (this is also called wildcard searching). This kind of filtering requires you to use the DAX CONTAINSSTRING() function. An example makes this easier to understand (in the Client table this time):

```
Contains Name =
CALCULATE(SUM(Sales[SalePrice]), CONTAINSSTRING('Client'[ClientName],
"motor"))
```

This measure will return the total sales for any client whose name contains *motor*. The text that you are looking for can be anywhere inside the field - or even be the entire field contents.

Should you wish for the "mirror image" if this output (that is all clients except those with motor somewhere in the name) you can add the NOT operator to the filter like this:

```
Does not Contain Name =
CALCULATE(SUM(Sales[SalePrice]),
NOT CONTAINSSTRING('Client'[ClientName], "motor"))
```

You can see these measures in Figure 9-11.

FullYear	Contains Name	Does not Contains Name
2019	94500	929925
2020	329300	3454395
2021	997200	5674185
2022	277355	6648310
Total	**1698355**	**16706815**

Figure 9-11.   Partial text filtering using CALCULATE( )

NULL (Blank or Empty Cell) Handling

Measures can easily filter on empty cells in fields. This is known as NULL handling in the world of many programming languages. The following short piece of DAX finds the total for all sales where the TotalDiscount field does *not* contain a figure:

```
No Discount Sales =
CALCULATE(SUM(Sales[SalePrice]), ISBLANK('Sales'[TotalDiscount]))
```

Inversely, to find the total for all sales where the TotalDiscount field *does* contain a figure, you can simply add the NOT operator to the DAX as shown as follows:

```
Discount Sales =
CALCULATE(SUM(Sales[SalePrice]),
NOT ISBLANK('Sales'[TotalDiscount]))
```

You can see these measures in action in Figure 9-12.

FullYear	Discount Sales	No Discount Sales
2019	638135	386290
2020	1615160	2168535
2021	1843070	4828315
2022	1653070	5272595
Total	**5749435**	**12655735**

Figure 9-12. *NULL handling with CALCULATE( )*

You can find these measures in the Sales table of the sample files.

Using Multiple Filters

For a final filter example, imagine that you want to isolate the percentage of creditworthy dealer sales relative to dealer sales. You can do this by using the two measure calculations (UK Sales and CreditworthyUK Sales) in a single measure to obtain the desired result.

As you saw in detail how to create these calculations in the previous pages, I will only show you the formula here (you can look in the Sales table of the file PrestigeCarsDimensionalWithFilteredMeasures.pbix to find it):

```
Creditworthy UK Sales Percent = CALCULATE(SUM(Sales[SalePrice]),
Client[IsReseller]=TRUE(),Client[IsCreditRisk]=FALSE()) / CALCULATE(SUM
(Sales[SalePrice]),Client[IsReseller]=TRUE())
```

As you can see (and much as you would in Excel), you can combine functions - even complex filtered functions - in a single metric to deliver some powerful analysis. Using this measure in a simple table displaying sales by Make for 2020 gives a result like the one in Figure 9-13.

Make	SalePrice	Creditworthy UK Sales Percent
Alfa Romeo	£243,510.00	100%
Aston Martin	£5,102,755.00	84%
Austin	£64,500.00	100%
Bentley	£1,689,240.00	91%
BMW	£60,500.00	100%
Bugatti	£1,915,500.00	100%
Citroen	£101,080.00	100%
Delahaye	£107,000.00	100%
Delorean	£99,500.00	
Ferrari	£5,540,850.00	79%
Jaguar	£882,955.00	61%
Lagonda	£218,000.00	100%
Lamborghini	£1,727,150.00	100%
McLaren	£295,000.00	
Mercedes	£471,115.00	84%
Morgan	£18,500.00	100%
Noble	£206,900.00	100%
Peugeot	£21,045.00	40%
Porsche	£1,169,240.00	91%
Reliant	£1,900.00	
Rolls Royce	£1,637,000.00	84%
Trabant	£8,440.00	100%
Triumph	£396,270.00	82%
Total	**£21,977,950.00**	**87%**

Figure 9-13. *Using multiple filters in a measure*

Now that you have learned how to create filtered measures you have mastered the building blocks of an extremely powerful technique that you can adapt and extend in your own data models.

The Extent of Filtering in CALCULATE( )

Filtering in CALCULATE() is extremely "literal minded." It also *replaces* any filters on the field that you are filtering on. This can be disconcerting until it is fully understood.

Take a look at the following DAX (in the Sales table) that calculates the total sales for Ferraris.

```
Ferrari Sales =
CALCULATE(SUM(Sales[SalePrice]), 'Vehicle'[Make] = "Ferrari")
```

If you add this measure to a table of makes and colors, you get the output that you can see in Figure 9-14.

Make		
☐ Alfa Romeo		
■ Aston Martin		
■ Bentley		
■ Ferrari		
☐ Jaguar		
☐ Lamborghini		
☐ Porsche		
☐ Rolls Royce		
☐ Triumph		

Make	Total Sales	Ferrari Sales
Aston Martin	£1,191,470	£1,083,450
Bentley	£732,140	£1,083,450
Ferrari	£1,083,450	£1,083,450
Total	**£3,007,060**	**£1,083,450**

FullYear	
☐ 2018	
☐ 2019	
■ 2020	
☐ 2021	
☐ 2022	
☐ 2023	

Make	Color	Total Sales	Ferrari Sales
Aston Martin	Black	£62,080	£602,450
Aston Martin	British Racing Green	£152,000	£156,500
Aston Martin	Dark Purple	£114,600	
Aston Martin	Green	£190,500	
Aston Martin	Night Blue		£165,000
Aston Martin	Pink	£102,500	
Aston Martin	Red	£402,900	
Aston Martin	Silver	£166,890	£159,500
Bentley	Black	£99,500	£602,450
Bentley	Blue	£171,650	
Bentley	British Racing Green		£156,500
Bentley	Canary Yellow	£142,390	
Bentley	Night Blue		£165,000
Bentley	Pink	£133,450	
Bentley	Silver	£185,150	£159,500
Ferrari	Black	£602,450	£602,450
Ferrari	British Racing Green	£156,500	£156,500
Ferrari	Night Blue	£165,000	£165,000
Ferrari	Silver	£159,500	£159,500
Total		**£3,007,060**	**£1,083,450**

Figure 9-14. *Explicit filters in CALCULATE( ) overriding all implicit filters on a field*

This is standard DAX behavior. After all, the filter says "show total sales for Ferraris." So it does just that - it overrides any implicit filters on the Make field to show output for Ferraris only. What may be more of a surprise is that this measure is still filtered implicitly by any field *other* than make. So you can slice and dice the measure by any other field and it will accept the filtering - and show the total sales for Ferraris that map to the implicit filtering. This includes rows for color (shown in the table on the right) as well as the year slicer common to both tables.

Limits on CALCULATE( ) Filters

The filtering approaches that you have learned in this chapter cover a large range of filtering options that you may need to apply to basic measures. There are, of course, many other more complex filtering challenges that DAX can handle. However it is important to understand that a CALCULATE() filter cannot:

- Use a measure as the filter comparison element (on the right of the equals sign)

- Use a measure or an aggregation as the element to filter (on the left of the equals sign)

Indeed, if you try and use a measure or an aggregation instead of a hard-coded value or a field reference as the comparison element, as in this DAX example

```
Bad Measure =
CALCULATE(SUM(Sales[SalePrice]), 'Sales'[SalePrice] = [Gross Profit])
```

You will get the error message shown in Figure 9-15.

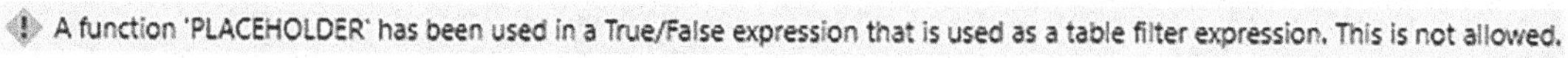

⚠ A function 'PLACEHOLDER' has been used in a True/False expression that is used as a table filter expression. This is not allowed.

Figure 9-15. *The Error message when using a measure as part of a filter in CALCULATE( )*

The same will happen if you try (for instance)

```
Bad Measure 2 =
CALCULATE(SUM(Sales[SalePrice]), [Gross Profit] = 50000)
```

This means that, for *simple* filtering with the CALCULATE() function, you are restricted to

- A column reference (on the left of the equals sign)

- A value (on the right of the equals sign)

This can be highly disconcerting for newcomers to DAX as it seemingly closes off extensive filtering opportunities. Yet there are solutions, as this limitation can be overcome using various DAX approaches. However, the solutions to these challenges require other DAX concepts to be introduced first. So, to make the DAX learning curve less steep, I will explain how to extend the use of the CALCULATE() function as various new concepts are introduced in the upcoming chapters.

Conclusion

This chapter introduced you to ways in which you can apply simple filters to measures. Although these are known as simple filters, they can apply logic, as well as filtering on text, numeric, Boolean, or date and time elements.

You also learned how to extend filters to apply slightly more complex filtering using multiple parameters and multiple elements in CALCULATE() filters.

There is a Power BI Desktop file that contains all the measures outlined in this chapter. It is the file PrestigeCarsDimensionalWithFilteredMeasures.Pbix that you can download from the sample data for this book on the Apress web site.

This chapter was only one further step on the path to learning how to apply measures. There is still plenty more to come. The next chapter will introduce you to overriding implicit filters when creating measures.

CHAPTER 10

CALCULATE() Modifiers

This chapter will continue explaining the CALCULATE() function. As I mentioned (and as you may have discovered if you have started writing DAX) CALCULATE() is the most complex - as well as the most powerful and probably the most useful - function available in DAX. So we need now to move on to the next level of understanding of what this function can deliver.

CALCULATE() can not only add explicit filters to formulas. It can also control how implicit filters interact with a calculation. This consists of overriding - or modifying if you prefer - the effects of implicit filters. This is a tremendously powerful feature that enables you to fine-tune how filters work. Be warned, however, that this function can be both complex to understand and complicated to implement.

In this chapter you will continue with the measures that you created in previous chapters. This means using the file PrestigeCarsDimensionalWithFilteredMeasures.Pbix.

Calculating Percentages of Totals

The filters that you have applied up until now in this chapter merely delivered subsets of data. Sometimes you will need filters to do the opposite, and apply a calculation to an entire data set - in effect overriding some or all of the implicit filters. In other words you will need filters that *remove* filters. This is often because calculating a total means telling DAX to aggregate a column without applying any of the filtering by row that would normally be applied. In other words you need to *prevent* the automatic filters that have proved so useful thus far.

239

© Adam Aspin 2023
A. Aspin, *Pro DAX and Data Modeling in Power BI*, https://doi.org/10.1007/978-1-4842-8995-2_10

A Simple Percentage

Imagine a table where you want to calculate the percentage of a total that each row represents. This could be the total of sales by Make, for instance. What you need to do here is to divide, fairly simply, the sales by the total sales. Here is how you can do this:

1. Click on the Sales table in the Fields List.

2. In the Modeling ribbon click on the New Measure button.

3. In the Formula Bar enter the following formula:

```
MakePercentage =
DIVIDE(SUM(Sales[SalePrice]),
CALCULATE(SUM(Sales[SalePrice]), REMOVEFILTERS(Vehicle[Make])))
```

If you create a simple table of sales per Make and add this new measure, it should look like Figure 10-1 (where the new measure output is formatted as a percentage for readability).

Make	SalePrice	MakePercentage
Alfa Romeo	£243,510.00	1.11%
Aston Martin	£5,102,755.00	23.22%
Austin	£64,500.00	0.29%
Bentley	£1,689,240.00	7.69%
BMW	£60,500.00	0.28%
Bugatti	£1,915,500.00	8.72%
Citroen	£101,080.00	0.46%
Delahaye	£107,000.00	0.49%
Delorean	£99,500.00	0.45%
Ferrari	£5,540,850.00	25.21%
Jaguar	£882,955.00	4.02%
Lagonda	£218,000.00	0.99%
Lamborghini	£1,727,150.00	7.86%
McLaren	£295,000.00	1.34%
Mercedes	£471,115.00	2.14%
Morgan	£18,500.00	0.08%
Noble	£206,900.00	0.94%
Peugeot	£21,045.00	0.10%
Porsche	£1,169,240.00	5.32%
Reliant	£1,900.00	0.01%
Rolls Royce	£1,637,000.00	7.45%
Trabant	£8,440.00	0.04%
Triumph	£396,270.00	1.80%
Total	**£21,977,950.00**	**100.00%**

Figure 10-1. Using the REMOVEFILTERS() function to calculate a percentage per attribute

This formula - and the concept behind it - may seem a little peculiar. So let me explain the REMOVEFILTERS() function in greater detail. In essence the REMOVEFILTERS() function says "remove all the filters concerning any specified fields." Consequently in this example the make is *not* filtered when calculating the total sales when it is used as the denominator of the DIVIDE() function. This means that the unfiltered total can now be calculated - and so can the percentage of each make relative to this grand total.

I can use this table as well to remind you of a fundamental principle of DAX, which is that each calculation is carried out individually. So the total percentage (100%) is *not* added up on the total row but is calculated by dividing the sum of the sales for the row by the sum of all sales. In other words the same formula is applied for every row in a table *including totals and subtotals*. This means that when the percentages of totals and subtotals are calculated, they too calculate the aggregate for the subtotal divided by the overall total. Remembering this is important when it will come to understanding other calculations a little later in this chapter.

It is important to note that the REMOVEFILTERS() function will *only remove filters for the fields that you have specified*. For instance, take a look at Figure 10-2, which shows a matrix for the sales and the percentage by make for two years.

FullYear	2020		2021		Total	
CountryName	Total Sales	Sales No Geography or Client	Total Sales	Sales No Geography or Client	Total Sales	Sales No Geography or Client
⊟ Belgium	£125,000	2.55%	£178,950	2.27%	£303,950	2.38%
False	£125,000	2.55%	£178,950	2.27%	£303,950	2.38%
⊟ France	£784,345	15.99%	£2,866,725	36.43%	£3,651,070	28.58%
False	£595,765	12.15%	£2,693,675	34.24%	£3,289,440	25.75%
True	£188,580	3.84%	£173,050	2.20%	£361,630	2.83%
⊟ Germany	£277,390	5.65%	£473,100	6.01%	£750,490	5.88%
False	£277,390	5.65%	£473,100	6.01%	£750,490	5.88%
⊟ Italy	£181,540	3.70%	£439,400	5.58%	£620,940	4.86%
False	£181,540	3.70%	£439,400	5.58%	£620,940	4.86%
⊟ Spain	£332,750	6.78%	£571,780	7.27%	£904,530	7.08%
False	£332,750	6.78%	£571,780	7.27%	£904,530	7.08%
⊟ Switzerland	£335,950	6.85%	£207,200	2.63%	£543,150	4.25%
False	£240,000	4.89%	£167,350	2.13%	£407,350	3.19%
True	£95,950	1.96%	£39,850	0.51%	£135,800	1.06%
⊟ United Kingdom	£2,574,010	52.47%	£2,759,070	35.07%	£5,333,080	41.75%
False	£2,289,910	46.68%	£2,567,420	32.63%	£4,857,330	38.03%
True	£284,100	5.79%	£191,650	2.44%	£475,750	3.72%
⊟ United States	£294,300	6.00%	£371,850	4.73%	£666,150	5.22%
False	£294,300	6.00%	£371,850	4.73%	£666,150	5.22%
Total	£4,905,285	100.00%	£7,868,075	100.00%	£12,773,360	100.00%

Figure 10-2. *The REMOVEFILTERS() function lets all other filters be applied*

You can see here that the measure MakePercentage is correctly applied independently to Sales for 2019, 2020, 2021, and 2022 as well as the grand total. This is because all other filters (the year in this case) *are* applied (as you have come to expect with Power BI Desktop), and *only* the Make is not filtered when calculating the total sales as it is the *only* field that is not deliberately overridden.

REMOVEFILTERS( ) or ALL( )?

If you inherit legacy DAX in older Power BI Desktop files or if you look at most of the web pages and books that explain how to tweak the filtering applied by the CALCULATE() function, you will probably see that instead of the REMOVEFILTERS() function, the ALL() function is applied.

So the previous formula could have been written

```
MakePercentage =
DIVIDE(SUM(Sales[SalePrice]),
CALCULATE(SUM(Sales[SalePrice]), ALL(Vehicle[Make])))
```

This formula would return exactly the same results as the REMOVEFILTERS() version. Indeed, the two can be used interchangeably when "unfiltering" a CALCULATE() parameter. However, the ALL() function (which is one of the oldest DAX functions) has many other uses. This means that it can get confusing trying to remember or work out which aspect of ALL() you are trying to apply to a formula.

So I prefer to use the REMOVEFILTERS() function when developing DAX CALCULATE functions that require modifiers. I find that "the clue is in the title" when using the word *removefilters* - and that this helps understand what you are doing to a formula. You can, of course, use ALL() if you prefer.

Removing Multiple Filter Elements

If your visualization is more complex than the simple example that you just saw, then you will have to craft your measures appropriately to handle any complexity. Take the case, for instance, where you want to see sales by Make *and* Color and display the percentage of each row compared to the total. You will need to calculate a total that discards the filters for Make *and* Color so that DAX can arrive at the correct figure for the overall total. Here is the formula that can do this:

```
MakeAndColorPercentage =
DIVIDE(
SUM(Sales[SalePrice]), CALCULATE(SUM(Sales[SalePrice])
, REMOVEFILTERS(Vehicle[Make]), REMOVEFILTERS(Vehicle[Color])
)
)
```

This code snippet shows that the CALCULATE() function can take multiple elements to "unfilter" a calculation just as it can accept multiple filters. The key is in the following part of the measure:

```
REMOVEFILTERS(Vehicle[Make]), REMOVEFILTERS(Vehicle[Color])
```

This piece of DAX is simply saying "don't apply any Make or Color filters when calculating." The consequence is that this measure now calculates the total sales whatever the Make or Color. So, once again, CALCULATE() consists of an initial aggregation followed by any number of parameters that alter the scope of the initial calculation. These parameters are called *modifiers* when they are not actually filtering data.

This formula becomes the basis for the percentage calculation that you can see in Figure 10-3.

Make	Color	Total Sales	MakeAndColorPercentage
Alfa Romeo	Black	£93,125	0.42%
Alfa Romeo	Blue	£46,450	0.21%
Alfa Romeo	British Racing Green	£21,050	0.10%
Alfa Romeo	Dark Purple	£21,500	0.10%
Alfa Romeo	Green	£17,500	0.08%
Alfa Romeo	Night Blue	£35,190	0.16%
Alfa Romeo	Red	£8,695	0.04%
Aston Martin	Black	£1,439,560	6.55%
Aston Martin	Blue	£355,500	1.62%
Aston Martin	British Racing Green	£259,385	1.18%
Aston Martin	Canary Yellow	£266,350	1.21%
Aston Martin	Dark Purple	£114,600	0.52%
Aston Martin	Green	£599,590	2.73%
Aston Martin	Night Blue	£662,090	3.01%
Aston Martin	Pink	£263,100	1.20%
Aston Martin	Red	£402,900	1.83%
Aston Martin	Silver	£739,680	3.37%
Total		**£21,977,950**	**100.00%**

Figure 10-3. *Using the REMOVEFILTERS( ) function to calculate a percentage for two attributes*

What is vital to understand at this point is that these measures will be *tightly linked to the data that you are displaying* in the visuals that use them. To make this clearer, take a look in Figure 10-4 when you add the MakePercentage measure that you created earlier.

Make	Color	Total Sales	MakeAndColorPercentage	MakePercentage
Aston Martin	Black	1439560	6.55%	21.50%
Aston Martin	Blue	355500	1.62%	13.76%
Aston Martin	British Racing Green	259385	1.18%	11.72%
Aston Martin	Canary Yellow	266350	1.21%	42.09%
Aston Martin	Dark Purple	114600	0.52%	9.84%
Aston Martin	Green	599590	2.73%	32.55%
Aston Martin	Night Blue	662090	3.01%	43.51%
Aston Martin	Pink	263100	1.20%	54.56%
Aston Martin	Red	402900	1.83%	18.41%
Aston Martin	Silver	739680	3.37%	27.90%
Bentley	Black	199450	0.91%	2.98%
Bentley	Blue	361150	1.64%	13.98%
Bentley	British Racing Green	153000	0.70%	6.91%
Bentley	Canary Yellow	142390	0.65%	22.50%
Bentley	Dark Purple	89500	0.41%	7.68%
Bentley	Night Blue	156450	0.71%	10.28%
Bentley	Pink	133450	0.61%	27.67%
Bentley	Red	112400	0.51%	5.14%
Bentley	Silver	341450	1.55%	12.88%

Figure 10-4. Understanding how modified measures depend on the output structure

The MakePercentage measure that worked beautifully in Figure 10-1 gives inaccurate figures when a table displays data by Make and color. A cursory glance at the make percentages tells you that the total for the percentage per make is way over 100%. Indeed it is over 100% for each make. This slightly disconcerting effect is because we are ignoring Make filters and what we are calculating here is a particular make's proportion of sales for that color car compared to the total sales for that color car.

It follows that you will need to be aware of how you want to apply a formula to a visual before you create it. So be prepared to take a close look at the rows and columns (for tables and matrices) or axes (for charts) before you dive into complex DAX formulas using CALCULATE() modifiers. The corollary is that you may well have to create multiple measures that carry out the same calculations in slightly different circumstances. These variations on a theme will probably be based on slightly different CALCULATE() modifiers that you use in different visuals.

REMOVEFILTERS( ) Constraints

While the REMOVEFILTERS() function opens up vast horizons when it comes to overriding the automated implicit filtering that Power BI applies by default, it does suffer from a kind of "tunnel vision." That is, you will need to be extremely precise when defining exactly which filters you wish to remove from a calculation.

This is best grasped with an example. If you take the table shown in Figure 10-1 and swap the CountryName field for the Make field in the MakePercentage measure, you get the output shown below in Figure 10-5.

CountryName	Total Sales	MakePercentage	CountryPercentage
Belgium	311850	100.00%	1.42%
France	5882810	100.00%	26.77%
Germany	873040	100.00%	3.97%
Italy	1170780	100.00%	5.33%
Spain	1693060	100.00%	7.70%
Switzerland	839415	100.00%	3.82%
United Kingdom	9211375	100.00%	41.91%
United States	1995620	100.00%	9.08%

Figure 10-5. *REMOVEFILTERS( ) is highly specific in its scope*

The point to note is that REMOVEFILTERS( ) is extremely literal-minded. This particular measure says "don't filter on make" in the percentage calculation. So that is what it does. No one mentioned the possibility of needing to see percentages of countries (and so to calculate the total for country sales and use this for a ratio). So DAX lets you see that the MakePercentage measure cannot be used with countries.

Fortunately it is easy to clone the existing MakePercentage formula and remove a different filter (I chose to add this to the Sales table):

```
CountryPercentage =
DIVIDE(SUM(Sales[SalePrice]),
CALCULATE(SUM(Sales[SalePrice]), REMOVEFILTERS('Geography'[CountryName])))
```

Figure 10-5 shows the new measure alongside the first one that you created in a table listing sales by country.

The key takeaway from this analysis (and at risk of belaboring the point) is that measures will need to be highly customized depending on the structure of the visual that you will be using. However, this does not necessarily mean that you will need hundreds of measures in your dashboards. The art when creating measures is to extend them so that they can accommodate a series of requirements.

Note   I advise against attempting to create overly complex "one size fits all" measures that can handle a multitude of use cases. These rapidly (and inevitably) become extremely complex and difficult to maintain. Best practice is nearly always to have groups of measures using a clear and coherent naming convention.

Extending the Scope of REMOVEFILTERS( )

While it is possible to create a measure for every single reporting requirement, this approach can lead to a data model being swamped in measures. It also tends to leave DAX developers and end users forgetting which is the appropriate measure among a vast selection that you should apply to solve a specific challenge. Consequently, it is nearly always more productive to create fewer measures where each can be applied to a wider range of dashboarding needs. However, the counter argument here is that it is never good practice to attempt to cram as much logic as possible into a single measure - or it becomes incomprehensible. So a measure should have a well-defined purpose - if that includes multiple bits of logic, then great, but only as long as they're sharing a common goal.

This can be achieved to a greater or lesser extent by extending the scope of how filters are removed inside the CALCULATE() function. This covers

- Defining a series of fields not to filter, one by one

- Specifying that all the fields in a table will not be filtered

- Removing filtering from an interconnected set of tables

- Specifying a small set of fields in a table that will remain filtered while all the other fields in the table are not filtered

Let's look at these in turn.

Defining a Series of Fields Not to Filter, One by One

As you have seen at the start of this chapter, the CALCULATE() function lets you specify any number of modifiers (or fields that should not be filtered if you prefer). All you have to do is to place them all in a comma-separated list after the first parameter of the CALCULATE() function. The example above could be extended like this:

```
Multi Unfilter =
CALCULATE(SUM('Sales'[DeliveryCharge]), REMOVEFILTERS(Vehicle[Make]),
REMOVEFILTERS(Vehicle[Color]), REMOVEFILTERS('Client'[ClientName]), REMOVEF
ILTERS('Geography'[CountryName])
)
```

As you can see from this simple DAX snippet each CALCULATE() parameter that acts as a modifier can refer to any table and field combination from the data model. I am placing this measure in the SalesInfo table; this time to show that measures can be added to any table.

Note   Overriding (or applying modifiers) to CALCULATE() presumes that the tables inside the parameters are part of a coherent data model. Specifically the filter direction must allow CALCULATE() modifiers to filter the data in the table containing the field that is aggregated.

Specifying That All the Fields in a Table Will Not Be Filtered

Specifying multiple fields in a table inside a CALCULATE() parameter can get laborious. So it is worth knowing that you can refer to the entire table inside a REMOVEFILTERS() function. This allows you to prevent any field from the specified table filtering the calculation. You can see this in the following DAX snippet that I am adding to the Geography table:

```
Sales No Geography =
DIVIDE(SUM(Sales[SalePrice]),
CALCULATE(SUM(Sales[SalePrice]), REMOVEFILTERS('Geography')))
```

Using this measure in a table that is based in the CountryName field, you will obtain the result shown in Figure 10-6.

CountryName	Total Sales	CountryPercentage	Sales No Geography
Belgium	311850	1.42%	1.42%
France	5882810	26.77%	26.77%
Germany	873040	3.97%	3.97%
Italy	1170780	5.33%	5.33%
Spain	1693060	7.70%	7.70%
Switzerland	839415	3.82%	3.82%
United Kingdom	9211375	41.91%	41.91%
United States	1995620	9.08%	9.08%
Total	**21977950**	**100.00%**	**100.00%**

Figure 10-6. *Removing all filters from the fields in a table*

If you swap out CountryName (from the Geography table) in the table visual and replace it with Town (also from the Geography table), you will get the result that you can see in Figure 10-7.

Town	Total Sales	CountryPercentage	Sales No Geography
Barcelona	948340	100.00%	4.31%
Berlin	184050	100.00%	0.84%
Birmingham	2557080	100.00%	11.63%
Brussels	311850	100.00%	1.42%
Geneva	371900	100.00%	1.69%
Lausanne	467515	100.00%	2.13%
Liverpool	1302115	100.00%	5.92%
London	2671115	100.00%	12.15%
Los Angeles	1087690	100.00%	4.95%
Lyon	256150	100.00%	1.17%
Madrid	744720	100.00%	3.39%
Manchester	959995	100.00%	4.37%
Marseille	3125225	100.00%	14.22%
Milan	704290	100.00%	3.20%
New York	485240	100.00%	2.21%
Newcastle	599430	100.00%	2.73%
Paris	2501435	100.00%	11.38%
Rome	466490	100.00%	2.12%
Stoke	1121640	100.00%	5.10%
Stuttgart	688990	100.00%	3.13%
Washington	422690	100.00%	1.92%
Total	**21977950**	**100.00%**	**100.00%**

Figure 10-7. *REMOVEFILTERS( ) displaying output from two tables*

So, as the comparison between these two tables shows, if you want a measure that can be used in a wider range of visuals (but always overriding filter fields from the same table), the REMOVEFILTERS() function applied to a table can help you to create more reusable measures for your data model.

This does not mean that you should push the logic to extremes and create measures that remove multiple filters "in case." You need to remember that most measures will need to allow certain implicit filters to function, while preventing others from being applied.

As an example, try adding the following DAX to the Sales table as a measure.

```
Sales No Geography or Client % =
DIVIDE(SUM(Sales[SalePrice]),
CALCULATE(SUM(Sales[SalePrice]),
        REMOVEFILTERS('Geography'),
        REMOVEFILTERS('Client'))
    )
```

If you look at Figure 10-8, you can see that *any combination of fields* from the Geography and Client tables can be used in conjunction with the Sales No Geography or Client measure to calculate the correct percentages. However the years used in the column headers apply the implicit filter for the year (it is not overridden in the measure) which means that the calculation works on a year by year basis.

FullYear	2020		2021		Total	
CountryName	Total Sales	Sales No Geography or Client	Total Sales	Sales No Geography or Client	Total Sales	Sales No Geography or Client
⊟ Belgium	£125,000	2.55%	£178,950	2.27%	£303,950	2.38%
False	£125,000	2.55%	£178,950	2.27%	£303,950	2.38%
⊟ France	£784,345	15.99%	£2,866,725	36.43%	£3,651,070	28.58%
False	£595,765	12.15%	£2,693,675	34.24%	£3,289,440	25.75%
True	£188,580	3.84%	£173,050	2.20%	£361,630	2.83%
⊟ Germany	£277,390	5.65%	£473,100	6.01%	£750,490	5.88%
False	£277,390	5.65%	£473,100	6.01%	£750,490	5.88%
⊟ Italy	£181,540	3.70%	£439,400	5.58%	£620,940	4.86%
False	£181,540	3.70%	£439,400	5.58%	£620,940	4.86%
⊟ Spain	£332,750	6.78%	£571,780	7.27%	£904,530	7.08%
False	£332,750	6.78%	£571,780	7.27%	£904,530	7.08%
⊟ Switzerland	£335,950	6.85%	£207,200	2.63%	£543,150	4.25%
False	£240,000	4.89%	£167,350	2.13%	£407,350	3.19%
True	£95,950	1.96%	£39,850	0.51%	£135,800	1.06%
⊟ United Kingdom	£2,574,010	52.47%	£2,759,070	35.07%	£5,333,080	41.75%
False	£2,289,910	46.68%	£2,567,420	32.63%	£4,857,330	38.03%
True	£284,100	5.79%	£191,650	2.44%	£475,750	3.72%
⊟ United States	£294,300	6.00%	£371,850	4.73%	£666,150	5.22%
False	£294,300	6.00%	£371,850	4.73%	£666,150	5.22%
Total	£4,905,285	100.00%	£7,868,075	100.00%	£12,773,360	100.00%

Figure 10-8. REMOVEFILTERS() applied to all fields from two tables

Yet if you use only the fields from the tables that have REMOVEFILTERS() applied, you get the output you can see in Figure 10-9. The data here has a slicer applied to show only figures for 2020 and 2021.

| IsCreditRisk | False | | True | | Total | |
CountryName	Total Sales	Sales No Geography or Client %	Total Sales	Sales No Geography or Client %	Total Sales	Sales No Geography or Client %
Belgium	£303,950	2.38%			£303,950	2.38%
France	£3,289,440	25.75%	£361,630	2.83%	£3,651,070	28.58%
Germany	£750,490	5.88%			£750,490	5.88%
Italy	£620,940	4.86%			£620,940	4.86%
Spain	£904,530	7.08%			£904,530	7.08%
Switzerland	£407,350	3.19%	£135,800	1.06%	£543,150	4.25%
United Kingdom	£4,857,330	38.03%	£475,750	3.72%	£5,333,080	41.75%
United States	£666,150	5.22%			£666,150	5.22%
Total	£11,800,180	92.38%	£973,180	7.62%	£12,773,360	100.00%

Figure 10-9. *REMOVEFILTERS() applied to all fields used in a matrix*

In this second application of the *same measure,* the total percentages for IsCreditRisk do not add up to 100% *for each value* (except in the totals column) as the total percentages are calculated using the total filtered by IsCreditRisk. This is accepted behavior in DAX and proves the point that you always create CALCULATE() modifiers as a function of any implicit filters provided by a visual - as well as the output that you want to achieve.

Removing Filtering from an Interconnected Set of Tables

A final level of filter reduction is to harness the power of the data model and to filter on a fact table. Or, to put things more generically, to filter on *any* table (a destination table containing metrics) that is connected to another table that the data model allows to filter the destination table. Let's see this in action first, and then extend the explanation after you have seen the results.

Try adding to the Sales table the following measure that removes all filters from the Sales table.

```
Sales No Filters =
DIVIDE(SUM(Sales[SalePrice]),
CALCULATE(SUM(Sales[SalePrice]), REMOVEFILTERS('Sales')))
```

You can see this measure in action in Figure 10-10.

CountryName	Make	IsCreditRisk	Total Sales	Sales No Filters
Belgium	Alfa Romeo	False	£23,000	0.10%
Belgium	Aston Martin	False	£211,500	0.96%
Belgium	Noble	False	£29,500	0.13%
Belgium	Peugeot	False	£4,900	0.02%
Belgium	Triumph	False	£42,950	0.20%
France	Alfa Romeo	False	£24,000	0.11%
France	Aston Martin	False	£452,500	2.06%
France	Aston Martin	True	£105,580	0.48%
France	Austin	False	£6,000	0.03%
France	Bentley	False	£289,000	1.31%
France	Bentley	True	£72,500	0.33%
France	BMW	False	£55,000	0.25%
France	Bugatti	False	£930,500	4.23%
France	Citroen	False	£30,500	0.14%
Total			**£21,977,950**	**100.00%**

Figure 10-10. *REMOVEFILTERS() applied to a fact table*

Here you can see that fields from any of the tables connected to (and capable of filtering) the fact table are having their filters overridden in this measure. So the percentage displayed is always the percentage of the grand total.

Remember that in the current data model, the fact table can be filtered by several dimensions. These are Geography, SalesInfo, Client, Date, and Vehicle.

So, in fact, the SalesNoFilters measure will remove filters from *all* of the tables that can follow a connection path to the fact table and where the filter direction is not blocked (that is filtering can flow from the filtering table to the destination table). This gives you a measure that can be used in multiple circumstances to provide the percentage of the total sales whatever the hierarchy applied in a table or matrix.

However this means, by extension, that you cannot add filters or apply slicers to this table without filtering the grand total as well. So this is an "all or nothing" approach.

Specifying a Small Set of Fields in a Table That Will Remain Filtered While All the Other Fields in the Table Are Not Filtered

In practice you could find yourself having to write extremely targeted measures that need to remove filters from nearly all elements in a calculation but one or two. So, to save you writing long lists of REMOVEFILTERS() or ALL() functions, you can say "All but" a field using the ALLEXCEPT() function

As an example of this (although it is extremely simple), suppose that you want to see the percentages of sales grouped by a sub-classification. You know that you want to have Make as the main grouping element, but then you might want to use Color, Client, or even Model as the subgroup. So, to save you having to write a measure specifically for each of these combinations, you can write the following (and add it to the Vehicle table):

```
AllButMakePercentage =
DIVIDE(SUM(Sales[SalePrice]), CALCULATE(SUM(Sales[SalePrice]),
ALLEXCEPT(Vehicle, Vehicle[Make])))
```

If you use this measure in a matrix where Make is the leftmost column, you can then add subgroups using any other field to get the kind of output that is shown in Figure 10-11.

Make	Total Sales	AllButMakePercentage
⊟ **Alfa Romeo**	**243510**	**100.00%**
Black	93125	38.24%
Blue	46450	19.08%
British Racing Green	21050	8.64%
Dark Purple	21500	8.83%
Green	17500	7.19%
Night Blue	35190	14.45%
Red	8695	3.57%
⊟ **Aston Martin**	**5102755**	**100.00%**
Black	1439560	28.21%
Blue	355500	6.97%
British Racing Green	259385	5.08%
Canary Yellow	266350	5.22%
Dark Purple	114600	2.25%
Green	599590	11.75%
Night Blue	662090	12.98%
Pink	263100	5.16%
Red	402900	7.90%
Silver	739680	14.50%
⊞ **Austin**	**64500**	**100.00%**
⊞ **Bentley**	**1689240**	**100.00%**
⊞ **BMW**	**60500**	**100.00%**
⊞ **Bugatti**	**1915500**	**100.00%**
⊞ **Citroen**	**101080**	**100.00%**
⊞ **Delahaye**	**107000**	**100.00%**
⊞ **Delorean**	**99500**	**100.00%**
⊞ **Ferrari**	**5540850**	**100.00%**
⊞ **Jaguar**	**882955**	**100.00%**
Total	**21977950**	**100.00%**

Figure 10-11. *Using the ALLEXCEPT( ) function to calculate a percentage per attribute per group*

In this example other filters (color here) are applied, but *not* make. So you are displaying the percentage for each color compared to the aggregate total for the make.

> **Note** ALLEXCEPT() does what it says, and removes all filters except the one that you specify. This can have the effect of preventing other filters from working as you expect.

Visual Totals

Users - and this means the target audience for your reports and dashboards - do not like anomalies or apparent contradictions. So you have to be sure that the data that they see is visually coherent. This is especially true when displaying tables and matrices with subtotals and grand totals.

One technique that can help you here is to use the ALLSELECTED() function. This will only apply any filters that have been added (either at Report-, Page-, or Visualization-level or as slicers or cross-filters from other visuals) *without* you having to specify the fields that you do not want to filter as you did in the previous examples.

Here is a measure – SalesPercentage - that uses the ALLSELECTED() function as the filter for the CALCULATE() function that returns the total much like you did previously. You can find this in the Sales table in the sample data.

```
SalesPercentage =
DIVIDE(SUM(Sales[SalePrice]),
CALCULATE(SUM(Sales[SalePrice]), ALLSELECTED()))
```

If you take a look at Figure 10-12, you will see that the subtotals for the sales percentage (and indeed the grand total) are accurate, despite the fact that the country USA is selected in a slicer.

Make	Total Sales	MakePercentage	SalesPercentage
Alfa Romeo	51950	2.60%	2.60%
Aston Martin	717020	35.93%	35.93%
Austin	2500	0.13%	0.13%
Bentley	181650	9.10%	9.10%
Ferrari	604500	30.29%	30.29%
Jaguar	120000	6.01%	6.01%
Mercedes	62100	3.11%	3.11%
Porsche	58400	2.93%	2.93%
Rolls Royce	182500	9.15%	9.15%
Triumph	15000	0.75%	0.75%
Total	**1995620**	**100.00%**	**100.00%**

Make	Total Sales	MakeAndColorPercentage	SalesPercentage
Alfa Romeo	**51950**	**2.60%**	**2.60%**
British Racing Green	12500	0.63%	0.63%
Dark Purple	21500	1.08%	1.08%
Night Blue	17950	0.90%	0.90%
Aston Martin	**717020**	**35.93%**	**35.93%**
Black	350530	17.56%	17.56%
Blue	91500	4.59%	4.59%
Green	49500	2.48%	2.48%
Night Blue	113590	5.69%	5.69%
Silver	111900	5.61%	5.61%
Austin	**2500**	**0.13%**	**0.13%**
Red	2500	0.13%	0.13%
Bentley	**181650**	**9.10%**	**9.10%**
Blue	92150	4.62%	4.62%
Dark Purple	89500	4.48%	4.48%
Total	**1995620**	**100.00%**	**100.00%**

Figure 10-12. *Using the ALLSELECTED() function to calculate a percentage per attribute per group*

The ALLSELECTED() function can make creating percentage totals much easier, as this function removes the need to create highly specific measures that are tied to specific types of calculation. As you can see (this is why I left the original measures MakePercentage and MakeAnd ModelPercentage in the output), the single measure SalesPercentage can be used anywhere that you want to display percentages as part of subtotals and grand totals. So you can use any different fields in Power BI tables and matrices in any combination and one formula – SalesPercentage - will give the percentage of sales to the total displayed in the table or matrix.

Explicit Measure Filters and Modifiers Cannot Be Overridden

You can use just about any measure inside another measure. So a measure that contains modifiers can be used as the first parameter of a second measure. This causes upstream changes (that is any changes to the first measure) to cascade down the calculation chain into the second - and indeed any subsequent - measures.

However you need to be aware that adding a modifier at a *second* level will *not* undo the effects of the filter or modifier at a *first* level.

Take a look at the following two short DAX snippets that are in the Vehicle table.

- The first adds a filter to a measure that already contains a filter (the "Ferrari Sales" measure only shows sales for Ferraris).

- The second explicitly removes all filters on the Vehicle table. This table contains the Make field, and so the measure removes all filters from the make field (including Ferraris).

```
Silver  Sales =
CALCULATE([Ferrari Sales], 'Vehicle'[Color] = "Silver")
```

And

```
All Vehicle Sales =
CALCULATE([Ferrari Sales], REMOVEFILTERS('Vehicle'))
```

In Figure 10-13 you can see that the silver sales are, in fact, the output for silver Ferraris and the All Vehicle Sales display the total for all Ferraris. So the dependent measure is limited to filtering and/or overriding the input from the first measure only insofar as this does not remove the initial filtering. In other words a dependent measure can only build on a previous measure and not override it. Any implicit filters that are not overridden by the new measures work as they would normally.

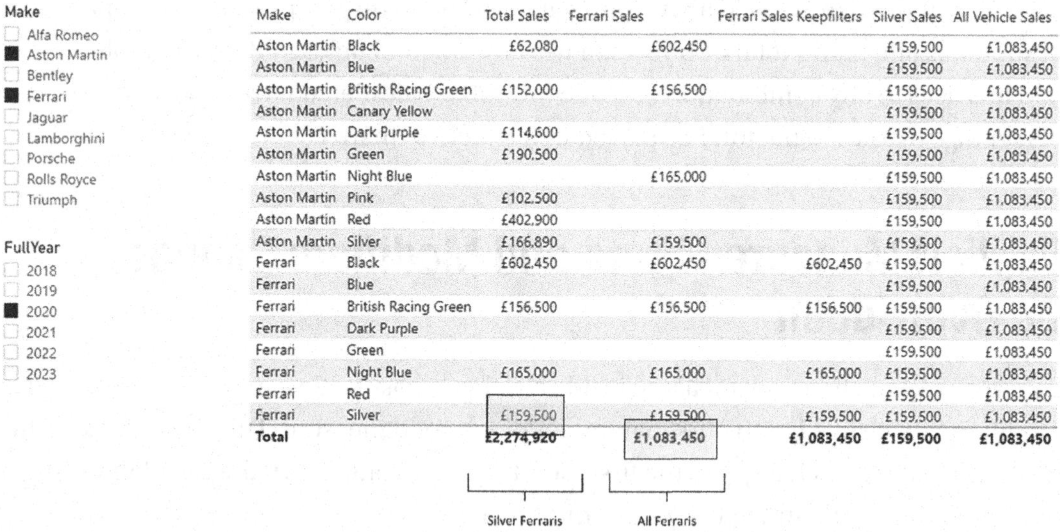

Make
- ☐ Alfa Romeo
- ■ Aston Martin
- ☐ Bentley
- ■ Ferrari
- ☐ Jaguar
- ☐ Lamborghini
- ☐ Porsche
- ☐ Rolls Royce
- ☐ Triumph

FullYear
- ☐ 2018
- ☐ 2019
- ■ 2020
- ☐ 2021
- ☐ 2022
- ☐ 2023

Make	Color	Total Sales	Ferrari Sales	Ferrari Sales Keepfilters	Silver Sales	All Vehicle Sales
Aston Martin	Black	£62,080	£602,450		£159,500	£1,083,450
Aston Martin	Blue				£159,500	£1,083,450
Aston Martin	British Racing Green	£152,000	£156,500		£159,500	£1,083,450
Aston Martin	Canary Yellow				£159,500	£1,083,450
Aston Martin	Dark Purple	£114,600			£159,500	£1,083,450
Aston Martin	Green	£190,500			£159,500	£1,083,450
Aston Martin	Night Blue		£165,000		£159,500	£1,083,450
Aston Martin	Pink	£102,500			£159,500	£1,083,450
Aston Martin	Red	£402,900			£159,500	£1,083,450
Aston Martin	Silver	£166,890	£159,500		£159,500	£1,083,450
Ferrari	Black	£602,450	£602,450	£602,450	£159,500	£1,083,450
Ferrari	Blue				£159,500	£1,083,450
Ferrari	British Racing Green	£156,500	£156,500	£156,500	£159,500	£1,083,450
Ferrari	Dark Purple				£159,500	£1,083,450
Ferrari	Green				£159,500	£1,083,450
Ferrari	Night Blue	£165,000	£165,000	£165,000	£159,500	£1,083,450
Ferrari	Red				£159,500	£1,083,450
Ferrari	Silver	£159,500	£159,500	£159,500	£159,500	£1,083,450
Total		£2,274,920	£1,083,450	£1,083,450	£159,500	£1,083,450

Silver Ferraris   All Ferraris

Figure 10-13. Attempting to override a filter and modifier in a dependent measure

KEEPFILTERS( )

A modified measure lives in an isolated "boxed off" universe. Through adding a modifier to a CALCULATE() function, you have ensured that the measure lives apart from the influence of any filters and slicers that concern the specific filter field(s) that have been overridden. If you need a visual reminder of this, then refer back to Figure 10-13.

Sometimes you need filters and slicers on the overridden filter to be applied nevertheless. This is where the KEEPFILTERS() function comes in. This function can be useful

- To create highly targeted measure calculations

- To develop a measure that is more reusable in a wider set of circumstances

In the previous chapter you saw how to create a measure "Ferrari Sales" that showed the sales for the best-selling make that the company sells. This gave the output that you saw in Figure 10-13 – and that you can see again in Figure 10-14.

Make	Color	Total Sales	Ferrari Sales	Ferrari Sales Keepfilters
Aston Martin	Black	£62,080	£602,450	
Aston Martin	British Racing Green	£152,000	£156,500	
Aston Martin	Dark Purple	£114,600		
Aston Martin	Green	£190,500		
Aston Martin	Night Blue		£165,000	
Aston Martin	Pink	£102,500		
Aston Martin	Red	£402,900		
Aston Martin	Silver	£166,890	£159,500	
Bentley	Black	£99,500	£602,450	
Bentley	Blue	£171,650		
Bentley	British Racing Green		£156,500	
Bentley	Canary Yellow	£142,390		
Bentley	Night Blue		£165,000	
Bentley	Pink	£133,450		
Bentley	Silver	£185,150	£159,500	
Ferrari	Black	£602,450	£602,450	£602,450
Ferrari	British Racing Green	£156,500	£156,500	£156,500
Ferrari	Night Blue	£165,000	£165,000	£165,000
Ferrari	Silver	£159,500	£159,500	£159,500
Jaguar	Black	£102,390	£602,450	
Jaguar	British Racing Green	£35,250	£156,500	
Jaguar	Canary Yellow	£99,500		
Jaguar	Night Blue	£19,600	£165,000	
Jaguar	Red	£39,600		
Jaguar	Silver	£3,575	£159,500	
Total		**£3,306,975**	**£1,083,450**	**£1,083,450**

Make: Alfa Romeo, ■ Aston Martin, ■ Bentley, ■ Ferrari, ■ Jaguar, Lamborghini, Porsche, Rolls Royce, Triumph

FullYear: 2018, 2019, ■ 2020, 2021, 2022, 2023

Output Using Keepfilters

Figure 10-14. *Applying the KEEPFILTERS() function*

Should you only want (for whatever reason) to ensure that this measure only appears when the make is the one used in the CALCULATE() filter (Ferrari in this example), then you can write DAX like the following.

```
Ferrari Sales Keepfilters =
CALCULATE(SUM(Sales[SalePrice]),
KEEPFILTERS('Vehicle'[Make] = "Ferrari")
)
```

You can see the output from this measure (added to the Sales table) alongside the output from the measure without KEEPFILTERS() in Figure 10-14.

In essence, you are now adding modifiers to modifiers. What this measure is saying is "calculate the total sales for Ferraris but don't completely remove the filter on the column where Ferrari is found." Consequently the Make filter is reapplied in the output – filtered on Ferrari. So the data is calculated only for Ferraris and is zero - because excluded by the filter - for all other makes.

Conclusion

This chapter took you further into the power and complexity of CALCULATE() by introducing you to modifiers.

CALCULATE() modifiers are parameters that allow you to specify that certain fields or sets of fields from one or more tables will not be affected by any implicit filtering. This can prevent rows, columns, chart axes, filters, and slicers from affecting a calculation.

You saw that there are a handful of modifiers that all are useful in specific circumstances. They are ALL(), REMOVEFILTERS(), ALLEXCEPT(), and ALLSELECTED().

Finally you learned about the KEEPFILTERS() function so that you can override modifiers to meet certain requirements.

The measures creates in this chapter are available in the file PrestigeCarsDimensionalWithModifiers.pbix. This file is in the sample data that is available for download from the Apress website.

CHAPTER 11

The Filter() Function

In Chapter 9 we have looked at filtering data using (fairly) simple comparisons applied as parameters to the CALCULATE() function. These kinds of data filters are both simple and efficient, but they have limits on what they can achieve. This chapter will push the boundaries of what can be done to select and subset data through introducing you to the FILTER() function.

It may help to consider the FILTER() function as a "helper" or "auxiliary" function. That is, you rarely - if ever - use it as the principle function in a DAX calculation. However you will probably find yourself using it repeatedly inside other functions to extend their capabilities and to deliver the results that you require.

FILTER() can be used in a vast array of ways inside a range of other DAX functions. However, so as not to swamp you with options, I will introduce FILTER() only when used with the CALCULATE() function at first. In later chapters you will extend your knowledge to apply FILTER() in other ways and using different approaches and functions.

This chapter will continue building on the techniques that you learned in the previous chapters. So you will need to open the Power BI Desktop file PrestigeCarsDimensionalWithModerators.pbix to develop the DAX measures that are explained in this chapter. This file is available to download from the Apress website.

Filter

FILTER() works by taking a table and then defining how the table will output fewer rows using a filter criterion. This means that FILTER() takes the following two parameters (in this order)

- A table
- Filter criteria

© Adam Aspin 2023
A. Aspin, *Pro DAX and Data Modeling in Power BI*, https://doi.org/10.1007/978-1-4842-8995-2_11

As ever, it is inevitably easier to understand more intuitively how the FILTER() function works using an example. Look at the following short DAX snippet in Figure 11-1.

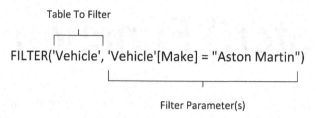

Table To Filter

FILTER('Vehicle', 'Vehicle'[Make] = "Aston Martin")

Filter Parameter(s)

Figure 11-1. *The FILTER() function*

There are a couple of things I need to say about filters from the start:

- You cannot use a FILTER() function directly to start a measure or a calculated column. This is why the example here does not begin with Name =.

- You will use the FILTER() function inside other functions - most often measures - as you will see in a couple of pages time.

- A FILTER() function creates a virtual table based on the source table defined as the first parameter and populates it with the set of rows specified by the second parameter.

- You will not normally see the result of the FILTER() function in a visual. Instead you will see the output of the primary measure that is enhanced by the FILTER() function.

Note   When I say that FILTER() takes a table as its first parameter, it can also take any DAX calculation or function that returns a table as its output. So "table" means any valid tabular DAX structure here.

Displaying the Output from a FILTER( )

The only initial difficulty with the FILTER() function is that you cannot see immediately any output that it delivers. There is, however, a simple technique that you can use to preview the output.

1. Switch to Data view using the Table icon on the top left of the Power BI Desktop window.

2. Click on any table to display the Table Tools menu.

3. In the Table Tools menu, click the New table button.

4. Enter the following short piece of DAX:

```
Aston Martins Only =
FILTER('Vehicle', 'Vehicle'[Make] = "Aston Martin")
```

5. Confirm the table function by pressing Enter or clicking the tick mark in the Formula bar.

A new table containing all the columns from the Vehicle table (but only rows containing Aston Martin) will be created looking like the one in Figure 11-2.

VehicleSK	Make	Model	Color	ModelVariant	MakeCountry	YearFirstProduced	YearLastProduced	VehicleType	IsRHD	Model (groups)
14	Aston Martin	DB2	Black		GBR			Coupe	1	DB2
15	Aston Martin	DB2	Black		GBR			Saloon	1	DB2
16	Aston Martin	DB2	Blue		GBR			Saloon	1	DB2
17	Aston Martin	DB2	British Racing Green		GBR			Saloon	1	DB2
18	Aston Martin	DB2	Canary Yellow		GBR			Saloon	0	DB2
19	Aston Martin	DB2	Green		GBR			Saloon	0	DB2
20	Aston Martin	DB2	Green		GBR			Saloon	1	DB2
21	Aston Martin	DB2	Night Blue		GBR			Saloon	1	DB2
22	Aston Martin	DB2	Red		GBR			Saloon	1	DB2
23	Aston Martin	DB2	Silver		GBR			Saloon	1	DB2

Figure 11-2. *Displaying the output of a FILTER() function as a table*

I need to stress that you will probably use FILTER() to *create* tables only very rarely. However I prefer to show you the application of a FILTER() function to create a table for a few reasons:

•   This shows you some concrete output from the FILTER() function. This, hopefully, makes the concepts more real.

- It makes the point that FILTER() returns a table. As you can see in Figure 11-2, the complete structure of the Vehicle table (the source table used in the first parameter) is maintained by the FILTER() function.

- It allows you to see that the filter criteria used as the second parameter have actually worked - only Aston Martin is the only make in the output.

- When developing complex formulas that involve FILTER() functions, it can help to test the output of the filter function this way. Of course, once you have verified that you are seeing the output that you need, you can delete the table as you would any other table that you want to remove from the data model.

Now that you have seen what a FILTER() function is, it makes sense to see how it actually works in real-world scenarios. Then we will move back to the actual function itself and take a closer look at more advanced filter parameters and complex filtering.

Filtering on Measures

The CALCULATE() function that we looked at in some detail in the previous three chapters is without a doubt one of the most powerful functions that you will use in DAX. However there are a few things that it *cannot* do. One of these is to filter data by comparing to a *measure* rather than to a column. If you cast your mind back to the examples where CALCULATE() was applied, you will remember that a data column or a calculated column was used every time that a comparison (text-based or numeric) was invoked. Indeed, if you try and use CALCULATE() with a measure rather than a column, you will get an error.

Fortunately DAX has a solution to this conundrum. It is to use the FILTER() function inside CALCULATE(). You may well wonder what differences there are, then, between using CALCULATE() to filter data and using the FILTER() function? Well, at its simplest FILTER() can use *measures* as part of a comparison, CALCULATE() must use columns - or calculated columns.

Let's see this in action. Suppose that you want to isolate sales where the ratio of gross profit is over 30%. Fortunately you have a measure - RatioGrossProfit - that calculates the percentage.

Just to make the point that you *have* to use the filter function, try the following approach first as a new measure in the Sales table using the RatioGrossProfit measure to filter the output:

```
High Net Margin Sales =
CALCULATE(SUM(Sales[SalePrice]), [RatioGrossProfit] > 0.3)
```

This formula will generate an error message rather than the output that we want. This is inescapable as CALCULATE() will simply *not* allow you to use a measure in a parameter.

Here is how you can use the Ratio Gross Profit measure with the help of a FILTER() function inside a CALCULATE() function so that you can display these lucrative sales. The formula should read

```
High Net Margin Sales =
CALCULATE(SUM(Sales[SalePrice]),
FILTER(Sales, [RatioGrossProfit] > 0.3))
```

If you use this measure in a table of sales by make and model, you should see something like Figure 11-3.

Make	Model	SalePrice	High Net Margin Sales
Alfa Romeo	1750	£13,525.00	
Alfa Romeo	Giulia	£89,695.00	£21,050
Alfa Romeo	Giulietta	£92,190.00	
Alfa Romeo	Spider	£48,100.00	
Aston Martin	DB2	£456,440.00	£62,500
Aston Martin	DB4	£281,700.00	£56,850
Aston Martin	DB5	£300,340.00	
Aston Martin	DB6	£1,019,095.00	£232,885
Aston Martin	DB9	£959,500.00	£124,000
Aston Martin	Rapide	£302,500.00	
Aston Martin	Vanquish	£361,040.00	
Aston Martin	Vantage	£343,500.00	£49,500
Aston Martin	Virage	£1,078,640.00	
Austin	Cambridge	£22,500.00	
Austin	Lichfield	£30,100.00	
Austin	Princess	£11,900.00	£2,250
Bentley	Arnage	£318,900.00	£76,500
Bentley	Brooklands	£289,000.00	
Bentley	Continental	£245,950.00	
Bentley	Flying Spur	£678,940.00	£218,890
Total		**£21,977,950.00**	**£2,846,340**

Figure 11-3. *Applying a filter to a measure*

In this example you have been able to filter data on a *measure* (Ratio Gross Profit) rather than a *column* inside the CALCULATE() function.

So this approach works. Yet I don't want just to be overjoyed that we have a working solution, but prefer to push a little further on with the reasons why this approach works.

Figure 11-4 shows the tabular output from the FILTER() function used inside the High Net Margin Sales measure. Remember that this DAX snippet used inside a measure was

```
FILTER(Sales, [Ratio Gross Profit] > 0.3)
```

| | 1 Virtual High Net Margin Filter Table = FILTER(Sales, [Ratio Gross Profit] > 0.3) |

SalePrice	CostPrice	TotalDiscount	DeliveryCharge	SpareParts	LaborCost	ClientSK	VehicleSK	GeographySK	DateSK	ID	Gross Profit
11500	7636		150	750	500	87	229	1	20190216	4	3864
6000	3984		150	750	500	27	4	11	20200217	43	2016
16500	10956		150	750	500	58	229	61	20200813	79	5544
35250	23406		550	750	500	26	160	30	20200914	102	11844
2250	1386		150	750	500	46	85	20	20210214	144	864
89500	59428		750	750	500	37	253	48	20210731	196	30072
1950	1295		150	750	500	33	209	10	20220105	282	655
2400	1478		150	750	500	42	217	53	20220730	294	922
2350	1560		150	750	500	40	216	27	20220730	296	790
8900	5910		150	750	500	39	289	7	20221102	336	2990
39500	24330		550	500	500	3	248	17	20201231	128	15170
990	657		150	150	500	36	114	38	20210331	155	333

Figure 11-4. *The output of the virtual filter table on gross profit*

As you can see when comparing the Gross Profit column with the Sale Price column, all the records in the virtual table returned by this specific FILTER() function meet the criterion of the gross profit being more that 30% of the sale price. So the filter is working.

The CALCULATE() function *takes the virtual table as its table parameter* - and only returns the sale price if the filter conditions are met. This means that the outer CALCULATE() function is only aggregating data on a subset of the Sales table. That is, the total sale price is returned if the Ratio Gross Profit measure is greater than 30%.

So you may find it easier to consider that the High Net Margin Sales measure is only carrying out the aggregation on a *subset* of the sales table. This subset is defined by the FILTER() function.

So now there are considerably fewer limitations on the use of CALCULATE() to develop much more complex outputs using measures as part of the filter parameter. CALCULATE() functions that use FILTER() - or even other approaches to go beyond the initial limitations - are generally referred to as complex CALCULATE() functions (as opposed to the simple CALCULATE() functions that you saw in Chapter 9 where you saw that CALCULATE can't take a measure in a filter expression).

Filter Criteria Inside the FILTER( ) Function

The filter criteria that you have seen used in a FILTER() function so far in this chapter have been extremely simple. You will doubtlessly need to create much more complex filters in your day-to-day work with DAX. So it is time to look at more intricate - and powerful - filter development.

Multiple Filter Conditions When Filtering on a Single Table

The first thing that I need to stress is that the filter parameter (or parameters) that you will use inside a FILTER() function is based on the filtering that you learned to apply when adding new (calculated) columns to a table. You may want to review the various filtering techniques that you learned in Chapter 9 before proceeding.

OR Filters

If the filter conditions that you want to apply are from the same table, you can use the AND/OR logic that you learned in Chapter 9.

So you could use the following approaches to filter parameters if you are filtering data on one or more parameters from inside the same table. By extension, filtering from elements in the same field is equally simple (and probably provoke a feeling of déjà vu if you have remembered all you learned in Chapter 9).

Anyway, take a look at the following applications of the FILTER() function.

```
FILTER('Geography', 'Geography'[CountryName] = "Spain" ||
'Geography'[CountryName] = "France")
```

Alternatively

```
FILTER('Geography', OR('Geography'[CountryName] = "Spain",
'Geography'[CountryName] = "France"))
```

Yet again

```
FILTER('Geography',  'Geography'[CountryName] IN {"Spain", "France"})
```

All of these variations on a theme allow you to filter on one or more elements from the CountryName field. As you saw example if output using these kinds of filter in Chapter 4 I will not repeat the output here.

Note   The FILTER() function does not need to be applied to the table where the measure is hosted. Once again, a measure is independent of the table where it is located.

AND Filters

To continue with basic filter parameter - and still from the same table - you can filter on one or more fields from the same table using DAX like this to combine filters to reduce the number of rows returned to the virtual table output form the FILTER() function.

```
FILTER('Vehicle', 'Vehicle'[Make] = "Ferrari" && 'Vehicle'[Color] = "Red")
```

Alternatively you can use

```
FILTER('Vehicle', AND('Vehicle'[Make] = "Ferrari", 'Vehicle'[Color]
= "Red"))
```

As a quick reminder, you can also use filter parameters to include all elements except a specific list. You can see this in the following DAX snippet and in Figure 11-5 the table that it creates.

```
FILTER('Vehicle', AND('Vehicle'[Make] = "Ferrari", NOT 'Vehicle'[Color] IN
{ "Red"}))
```

VehicleSK	Make	Model	Color	MakeCountry	VehicleType	IsRHD	Model (groups)
124	Ferrari	308	Black	ITA	Saloon	False	Other
125	Ferrari	355	Black	ITA	Coupe	False	Other
126	Ferrari	355	Black	ITA	Coupe	True	Other
127	Ferrari	355	Black	ITA	Saloon	True	Other
128	Ferrari	355	Blue	ITA	Saloon	True	Other
129	Ferrari	355	British Racing Green	ITA	Saloon	True	Other
131	Ferrari	355	Silver	ITA	Saloon	True	Other
132	Ferrari	360	Black	ITA	Saloon	True	Other
133	Ferrari	360	Blue	ITA	Saloon	True	Other
134	Ferrari	360	British Racing Green	ITA	Coupe	True	Other
135	Ferrari	360	Night Blue	ITA	Saloon	False	Other
136	Ferrari	Daytona	Black	ITA	Saloon	True	Other
137	Ferrari	Daytona	Silver	ITA	Saloon	True	Other
138	Ferrari	Dino	Black	ITA	Saloon	True	Other
139	Ferrari	Dino	British Racing Green	ITA	Saloon	True	Other
140	Ferrari	Dino	Dark Purple	ITA	Saloon	True	Other
141	Ferrari	Enzo	Black	ITA	Saloon	True	Other
142	Ferrari	Enzo	British Racing Green	ITA	Saloon	True	Other
143	Ferrari	F40	Black	ITA	Saloon	True	Other
144	Ferrari	F40	Dark Purple	ITA	Saloon	True	Other
145	Ferrari	F50	Blue	ITA	Saloon	True	Other
146	Ferrari	F50	Green	ITA	Coupe	True	Other
147	Ferrari	F50	Silver	ITA	Saloon	False	Other
148	Ferrari	Mondial	Black	ITA	Saloon	True	Other
149	Ferrari	Testarossa	Black	ITA	Saloon	True	Other
150	Ferrari	Testarossa	Green	ITA	Saloon	True	Other
151	Ferrari	Testarossa	Night Blue	ITA	Saloon	True	Other

No Red rows

Figure 11-5. *The output of the filter table using NOT … IN*

Strings

You can filter on the contents of a data element just as you can filter on the entire element. You can see this applied inside a FILTER() function in the following DAX snippet.

```
FILTER(Client, CONTAINSSTRING(Client[ClientName], "Wheel"))
```

Figure 11-6 shows the effect of applying a filter like this.

Figure 11-6. *The output of the filter table using NOT ... IN*

In case you are wondering, there are indeed three clients with the same name in the sample data. They are, however, in different locations.

Numbers

Filtering on numbers is, if anything, easier than filtering on text. The following snippet of DAX shows how to specify a numeric comparison inside a FILTER() function.

```
FILTER(Client, [CustomerID] > 10)
```

Specifying ranges is as easy as providing the lower and upper numeric thresholds like this

```
FILTER(Client, AND([CustomerID] > 10, [CustomerID] < 20))
```

Dates

Just as you can filter on numbers, you can filter on dates equally easily. The following piece of DAX shows how to specify a precise date inside a FILTER() function.

```
FILTER('SalesInfo', 'SalesInfo'[DateBought] >= VALUE("2019/01/01"))
```

For a range of dates, you can use DAX like this

```
FILTER('SalesInfo',
'SalesInfo'[DateBought] >= VALUE("2019/01/01") && 'SalesInfo'[DateBought]
<= VALUE("2019/12/31"))
```

Remember - and as explained in Chapter 7 - that you can also enter specific dates using the DATE() function.

Blank

Isolating empty fields in records can be done using the ISBLANK() function like this

```
FILTER('Vehicle', ISBLANK('Vehicle'[ModelVariant]))
```

Boolean

Finally, true/false filtering (or Boolean filters if you want to use the technical term) can be applied in any of the following ways:

```
FILTER('Client', 'Client'[IsCreditRisk])
```

An alternative is

```
FILTER('Client', 'Client'[IsCreditRisk] = TRUE())
```

Negating the true/false value can be done like this

```
FILTER('Client', NOT 'Client'[IsCreditRisk])
```

Or like this

```
FILTER('Client', 'Client'[IsCreditRisk] = FALSE())
```

Filtering on Criteria from Different Tables

FILTER() is not, however, a "magic bullet" when it comes to allowing you to filter anything anywhere anyhow. FILTER() has, in many ways, the same constraints as calculated columns in that you cannot apply filter parameters from tables other than the "main" table that you are using as the first parameter without using the RELATED() or RELATEDTABLE() functions. More specifically, applying filter parameters to FILTER() means respecting - and capitalizing on - the data model.

As an example of this, look at what happens when you want to filter a table based on the contents of a column in another table. This means using the RELATED() function, like this:

```
FILTER('Sales', RELATED('Geography'[CountryName]) = "United Kingdom")
```

There must, however, be a many to one relationship for this to work as you are looking up a single value in the Geography table from the Sales table. This was the case when adding new columns.

You can create more complex filters inside a FILTER() function by applying multiple filter parameters. The following DAX snippet illustrates this.

```
FILTER('Sales', RELATED(Vehicle[Make]) IN {"Rolls Royce", "Aston Martin",
"Ferrari"} && RELATED('Geography'[CountryName]) = "United Kingdom")
```

Equally, you can create complex AND filters by specifying several comma-separated filter parameters. Nesting FILTER() functions like this can allow some extremely powerful and specific filtering to be applied.

FILTER( ) Caveats

To conclude this short introduction to basic filtering with the FILTER() function, you need to be aware of a few potential limitations of this otherwise powerful DAX function.

- Be aware that the FILTER() function can be slow when applied to large datasets.

- Filter functions can become complex extremely quickly.

Filters, nonetheless, are a fundamental and extremely powerful part of your DAX armoury. There is still more that they can do, and they can be used in many contexts and situations. Consequently you will be exploring further aspects of the FILTER() function in later chapters to see how FILTER() can enable you to deliver more subtle and powerful analytics using DAX.

Conclusion

In this short chapter, you met the FILTER() function. This function extends the knowledge that you acquired previously when looking at filtering data in new columns to allow you to create virtual tables of filtered data that you can then use inside other functions such as CALCULATE().

You saw how to apply text, numeric, date, and Boolean filters as well as how to combine filter elements using AND/OR logic to deliver subsets of data. You also learned how to nest FILTER() functions to create more complex cumulative filters.

There is, however, more to FILTER() than we were able to cover in this chapter. So you will now see more ways of using FILTER() as you learn all about iterator functions in the next chapter.

Several of the functions that you created in this chapter are available in the file PrestigeCarsDimensionalWithFilter.Pbix that can be found with the sample data for this book.

CHAPTER 12

Iterators

In the previous chapter you discovered how to use the FILTER() function to reduce the scope of calculations. In fact, when you learned about FILTER(), you learned how to create an *iterator function. Iterators* - to give them their usual name - are a set of DAX functions that allow you to process dataset tables (or even the virtual tables that you will learn about in Chapter 16) and add calculations.

The remaining iterator functions are different from FILTER(). This is because FILTER() does not add any new columns or calculations but actually *reduces* the number of rows in the virtual table it returns. Other iterator functions add new columns to the virtual table that they create. Iterators are most often used in measures. You can then use their output in a variety of ways to deliver more powerful analytics in Power BI dashboards.

Iterators can be used to

- Add "virtual" calculated columns to tables that you can use just as you would the output from a calculated column

- Ensure that you are calculating output at a row level and not over an entire column

- Be used as "helper" functions inside other DAX calculations to allow the development of more advanced formulas by inputting ad-hoc calculations

The iterator functions that we will look at in this chapter extend the data model through adding new calculation elements. Although iterators work on a row-by-row basis, they are nonetheless measures.

All of this should become clear as you look at the examples in this chapter. Nonetheless I wanted to outline the principles first as they are fundamental to understanding what iterator functions are and how they work.

© Adam Aspin 2023
A. Aspin, *Pro DAX and Data Modeling in Power BI*, https://doi.org/10.1007/978-1-4842-8995-2_12

This chapter will also build on the dataset that you have developed in the previous chapters. This means downloading and using the Power BI Desktop file PrestigeCarsDimensionalWithFilter.Pbix.

DAX Iterator Functions to Replace Calculated Columns

In Chapter 4 you learned the basics of calculated columns. These can be such an easy solution that it is often a shame not to create them. However, they are stored in the table where they are added and take up space - on disk and in memory. What is more, adding calculated columns results in longer load times, particularly if tables containing calculated columns contain millions of rows.

In contrast, measures are only calculated at run time, and take up virtually no space. So, if you are considering creating many calculated columns, perhaps some of them could become measures instead. This can be done in certain circumstances using DAX iterator functions.

Let's see this in a practical example. In Chapter 4 you created a measure called Direct Costs. The DAX used was

```
DirectCosts = [CostPrice] + [SpareParts] + [LaborCost]
```

This column can be replaced by a more "lightweight" approach which is to use the DAX iterator SUMX(). You can see this iterator function, added to the Sales table, in the following DAX.

```
Total Costs =
SUMX(Sales, 'Sales'[CostPrice] +'Sales'[SpareParts] + 'Sales'[LaborCost])
```

This DAX iterator measure and the DAX for the calculated column give the same output as you can see in a table in Figure 12-1.

Make	Model	DirectCosts	Total Costs
Alfa Romeo	1750	£13,934	£13,934
Alfa Romeo	Giulia	£78,883	£78,883
Alfa Romeo	Giulietta	£83,086	£83,086
Alfa Romeo	Spider	£44,234	£44,234
Aston Martin	DB2	£356,967	£356,967
Aston Martin	DB4	£220,162	£220,162
Aston Martin	DB5	£252,708	£252,708
Aston Martin	DB6	£789,738	£789,738
Aston Martin	DB9	£749,173	£749,173
Aston Martin	Rapide	£237,986	£237,986
Aston Martin	Vanquish	£285,742	£285,742
Aston Martin	Vantage	£270,982	£270,982
Aston Martin	Virage	£871,584	£871,584
Austin	Cambridge	£20,470	£20,470
Austin	Lichfield	£25,315	£25,315
Austin	Princess	£14,064	£14,064
Bentley	Arnage	£245,940	£245,940
Bentley	Brooklands	£248,329	£248,329
Bentley	Continental	£192,400	£192,400
Bentley	Flying Spur	£523,769	£523,769
Total		**£17,738,835**	**£17,738,835**

Figure 12-1. SUMX() compared to a calculated column

What is more, both these DAX functions can be filtered and sliced in exactly the same way. The major difference is that, as it is a *measure* and *not* a new *column*, the Total Costs calculation can be added to *any table* in the data model. Moreover (at risk of belaboring the point) the measure is extremely lightweight and will not increase the size of the Power BI file or increase the load time in any perceptible fashion.

Iterator Parameters

All iterators have a fairly similar structure. They require at least two parameters which are

- A table to use as the data source

- A calculation to apply to the columns in the table

Optionally - and only in the case of a couple of iterator functions - other specific parameters may be required. You will see this towards the end of this chapter with the RANKX() function.

Figure 12-2 explains the standard iterator function structure in a more graphical way.

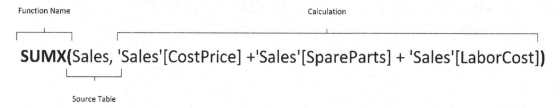

Figure 12-2. *The SUMX( ) function in DAX*

Iterators get their name from the fact that they iterate over the input table defined as the first parameter of the iterator function. So what is happening behind the scenes each time that you create an iterator is that the source table is being processed - or iterated *line by line* to be more precise - and the calculation that you specified is added for *each row individually*. However, nothing is physically added to the source table, and the calculation is only carried out when the iterator measure is used in a Power BI Desktop visual, or the visual is filtered, sliced, or otherwise updated.

Aggregator and Iterator Functions

Aggregator and iterator functions clearly have a lot in common. Yet there are fundamental differences. In essence (and without getting too abstruse too soon)

- Both return a single value (called scalar values or scalar output).

- Aggregators operate over an entire column (they work vertically).

- Iterators look at every row in the source table and apply the required calculation for each individual row (they work horizontally).

- Internally aggregation functions are converted to iterators (so SUM() is converted to SUMX()).The only difference is in syntax, which also gives the impression that one "works over a column" and the other "works over rows."

Much of the art of writing DAX that solves analytics challenges is to know when to aggregate and when to iterate.

Iterators and the Data Model

To make the point that iterators can be considered as a parallel version of calculated columns, let's look at how you could use an iterator to carry out a calculation that requires data from two or more tables.

Imagine that you want to see if sales have attained the minimum required (or that the sales manager hopes for). Prestige Cars has a simple formula for setting targets: each geographical area has a "risk factor" that is used to multiply the cost price of each vehicle to reach a target price. However, the SalePrice field is in the Sales table, whereas the RiskFactor field is in the Geography table. So you need, when using iterators, to create DAX like this (in the SalesInfo table to make the point that measures can be placed anywhere - even if we are iterating over the Sales table):

```
Risk Weighted Target Price =
SUMX(Sales, 'Sales'[CostPrice] * RELATED(Geography[RiskFactor]))
```

You can see the output from this formula in Figure 12-3. Here I have shown the output from the function in two tables at different levels of granularity so you can see that it respects the filtering, slicing, and level of detail simply and flawlessly.

ClientName	SalePrice	Risk Weighted Target Price
Alex McWhirter	£17,850.00	15,708.00
Alexei Tolstoi	£521,490.00	458,911.20
Alicia Almodovar	£382,090.00	336,239.20
Andrea Tarbuck	£454,500.00	399,960.00
Andy Cheshire	£174,500.00	153,560.00
Antonio Maura	£550,330.00	484,290.40
Autos Sportivos	£65,890.00	57,983.20
Beltway Prestige Driving	£54,875.00	48,290.00
Birmingham Executive Prestige Vehicles	£469,740.00	413,371.20
Bling Bling S.A.	£345,000.00	303,600.00
Bling Motors	£96,200.00	84,656.00
Boris Spry	£155,040.00	136,435.20
Bravissimat	£95,090.00	83,679.20
Capots Reluisants S.A.	£583,115.00	513,141.20
Casseroles Chromes	£201,050.00	176,924.00
Clubbing Cars	£57,900.00	50,952.00
Convertible Dreams	£361,995.00	318,555.60
Diplomatic Cars	£224,000.00	197,120.00
Total	**£21,977,950.00**	**19,810,744.80**

ClientName	SalesInfo_ID	SalePrice	Risk Weighted Target Price
Alex McWhirter	331	£17,850.00	15,708.00
Alexei Tolstoi	105	£45,950.00	40,436.00
Alexei Tolstoi	19	£22,990.00	20,231.20
Alexei Tolstoi	29	£22,600.00	19,888.00
Alexei Tolstoi	32	£69,500.00	61,160.00
Alexei Tolstoi	52	£9,950.00	8,756.00
Alexei Tolstoi	53	£39,500.00	34,760.00
Alexei Tolstoi	65	£205,000.00	180,400.00
Alexei Tolstoi	74	£102,500.00	90,200.00
Alexei Tolstoi	93	£3,500.00	3,080.00
Alicia Almodovar	114	£12,500.00	11,000.00
Alicia Almodovar	128	£39,500.00	34,760.00
Alicia Almodovar	165	£12,500.00	11,000.00
Alicia Almodovar	204	£250,000.00	220,000.00
Alicia Almodovar	252	£5,690.00	5,007.20
Alicia Almodovar	259	£15,950.00	14,036.00
Alicia Almodovar	266	£45,950.00	40,436.00
Andrea Tarbuck	250	£355,000.00	312,400.00
Total		**£21,977,950.00**	**19,810,744.80**

Figure 12-3. *Using RELATED( ) in iterators*

The takeaway here is that iterators work at row level just like calculated columns do. This means that, although you can attach them to any table, they look at the data model from the viewpoint of the table given as the first parameter.

In this example the Sales table is specified in the first parameter as the table to iterate over. Consequently, the iterator function acts like a new column in the Sales table, and can only retrieve data from other tables if

- There is a connection path between the table that is being iterated and the other table.

- RELATED() is used for a many (in the iterated table) to one (in the other table) lookup.

- RELATEDTABLE() is used for a one (in the iterated table) to many (in the other table) lookup.

- The cross-filter path allows the lookup.

As these concepts were explained in Chapter 4, I will not repeat them in detail here. However, you need to know that the same limitations and techniques are common to the two approaches - calculated columns and iterators.

Iterator Functions or Calculated Columns?

At this point you may well be asking yourself why we spent chapters on describing calculated columns when iterators can also do the job. Well, there are several reasons for this approach:

- Calculated columns are much easier to learn. What is more, they allow you to leverage your Excel skills and transition to DAX in a more measured and linear way, without having to jump straight into totally new and potentially strange concepts like table iteration.

- Learning calculated columns gives you a grounding in basic DAX that can then be further leveraged when you move into using (among other things) iterators.

- Calculated columns necessitate an understanding of core data model concepts such as filter direction and join paths.

- Iterators can be considered an extension of calculated columns in some respects. This makes iterators easier to understand once you are completely at ease with calculated columns.

- Iterators make for narrower tables that can will always load faster and take up less memory.

- Calculated columns are calculated once on refresh, so are static. Iterator measures are calculated on the fly, so react to the filters that are applied and are recalculated.

- When starting out with DAX, it can be easier to create multiple calculated columns where the output from one column becomes the input for another column. This can make creating complex column-based DAX easier.

- Sometimes an iterator is the only valid choice. This is the case with the RANKX() function that you will see later in this chapter.

AVERAGEX( ): The Ratio of the Sum vs. the Sum of the Ratio

There are cases when using an iterator can produce different results compared to using an aggregation function. Averages are a case in point.

More precisely, I want to outline a couple of ways to evaluate data on a row-by-row basis, yet return the result with any filters and slicers applied. There are many cases where this *cannot* be done using a calculated column and then returning the aggregate of the column data. After all, you need to return the *ratio of the sum* of any values and *not* the *sum of the ratio*. Think, for instance, of calculating a ratio for each row and then averaging the results to get the average ratio; it will be arithmetically false.

Power BI Desktop, fortunately, has some simple yet powerful solutions to this kind of conundrum. One principal tool is the use of a subset of the iterator functions that are aggregate functions. These include AVERAGEX(), COUNTX(), SUMX(), MAXX(), and MINX() among others. As you can see from the function names, they are the same names as the basic aggregation functions that you saw in Chapter 8 - but with an X added at the end of the function name.

These functions allow you to specify, as is the case with all iterators

- The table in which the calculations will apply

- The row-by-row calculation that is to be applied

As an example, consider the requirement for the ratio of the cost of any parts and labor compared to the sale price of each vehicle. Not only do we need this potentially at the finest level of granularity — the individual record — but we may need it sliced and diced by any number of criteria.

To extend your knowledge, we will also create this measure directly in the Report view, without stepping sideways into the Data view. So here is how to create a measure that you could use in Power BI Desktop reports:

```
Average Parts And LaborCost Ratio =
AVERAGEX(Sales,
DIVIDE('Sales'[SpareParts] + 'Sales'[LaborCost], 'Sales'[CostPrice])
)
```

1. Switch to Report View (unless this is already active).

2. In the Fields Pane, click on the ellipses to the right of the Sales table (or right-click on the Sales table) to display the context menu.

3. Select New Measure and add the DAX shown immediately above.

4. Confirm the measure creation.

Just creating the formula is pretty meaningless. So take a look at the chart in Figure 12-4, where you can see how this ratio instantly shows you which makes are the most costly (even to the extent of being loss-making) as a percentage of sales as far as spare parts and labor costs are concerned.

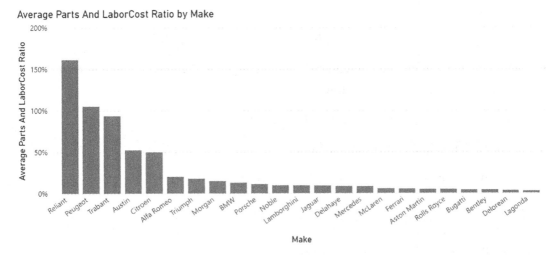

Figure 12-4. *Using the AverageX( ) function in DAX*

What this formula has done is add together the SpareParts and LaborCost, then divide by the SalePrice for *every row in the table*, and then return the average dependent on the filters and selections currently applied. This way, you will always get the mathematically accurate result in your visualizations.

The essential point to take away from this example is

- You can, and indeed, must, nest calculations inside parentheses to force Power BI Desktop to calculate elements in the correct order. This works in exactly the same way as it does in Excel, so I will not labor the point here.

Note   There is no direct way of displaying the iterator output as a new table as you saw in the previous chapter when using the FILTER( ) function. However, as iterators are Measures, the easy way to test the output is simply to add the measure to a visual (a card or a table, for instance) and check that the result is what you expect.

Filtering the Table Input for an Iterator

So far in this chapter you have always seen a table from the data model used as the first parameter for an iterator. This is not, however, compulsory. The first parameter can be

- A table

- The output of another DAX function that returns a table

What this boils down to is realizing that the first parameter can cover a wide range of options. For instance, it could contain

- The FILTER() function applied to a table

- DAX modifiers applied to a table

- A virtual table (which you will discover in Chapter 16)

Let's see the first two in action.

Using FILTER( ) Inside an Iterator

FILTER() is often used in conjunction with iterators in DAX to reduce the number of rows that will be iterated in the first parameter of the aggregator function.

As a first example, suppose that you need to create a measure (in the Sales table) that isolates the spare parts and labor cost where cheaper cars are concerned using DAX like this:

```
Sale Price for Cheap Sales =
SUMX(
FILTER(Sales, 'Sales'[SalePrice] < 50000),
'Sales'[SalePrice])
```

This DAX returns the sum of the sale price from the Sales table only where the sale price is more than 50000. As you can see, the filter is applied using the kind of approach that you have seen in Chapter 11.

Figure 12-5 shows the output for this measure compared to the total sale price whatever the retail value of any individual car. Here, the output is filtered on Aston Martin to make the point that measures such as these can be sliced "out of the box."

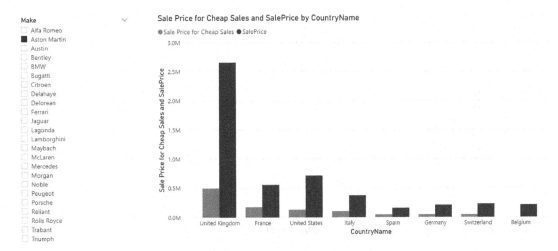

Figure 12-5. *Extending an iterator to filter the source data*

To show a slightly more complex FILTER() function in action, I will take the DAX function Average Parts and LaborCost Ratio that you created a few pages ago to calculate the ratio of average parts and labor cost to sales and extend it (as a new measure in the Sales table) to output this metric only for red cars.

The DAX to handle this particular requirement is

```
Average Parts And LaborCost Ratio For Red Cars =
AVERAGEX(
FILTER(Sales, RELATED('Vehicle'[Color]) = "Red"),
DIVIDE(('Sales'[SpareParts] + 'Sales'[LaborCost]), 'Sales'[CostPrice])
)
```

Figure 12-6 shows the average parts and labor cost ratio for red cars per make compared to the average parts and labor cost ratio for any color of car per make.

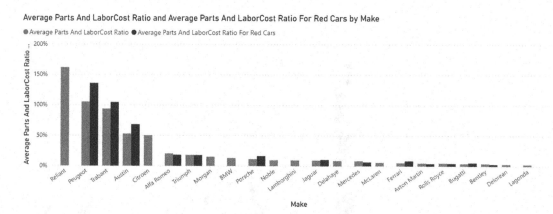

Figure 12-6. *Using FILTER( ) to extend an iterator*

To make this slightly more complex function more comprehensible, Figure 12-7 explains how it works in a more visual way.

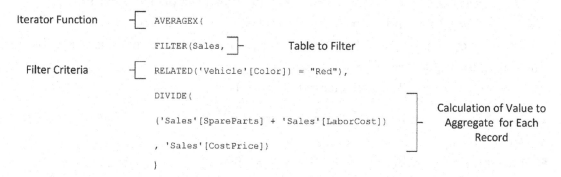

Figure 12-7. *Visual explanation of nesting FILTER( ) inside an iterator*

As is the case for all measures, these metrics can be sliced and diced in any number of ways.

Using Moderator Functions Inside an Iterator

The formula that shows the average parts and labor cost ratio will be filtered by anything that you choose to slice and dice it by. Yet there could be cases where you want to restrict how a slicer or filter interacts with an iterator.

Suppose, for instance, that you want to show how the sum of parts and labor costs relate to the grand total for parts and labor costs. This can be done by applying moderator functions to the first parameter of the iterator (the table).

To see this in action, create the following measure in the Sales table:

```
Ratio of SpareParts and Labor Cost to All Sales =
DIVIDE(
SUMX(
Sales,
('Sales'[SpareParts] + 'Sales'[LaborCost])
)
,SUMX(
ALL(Sales),
('Sales'[SpareParts] + 'Sales'[LaborCost])
)
)
```

The moderator functions (or moderators if you prefer) that you can use in cases like these are

- ALL()

- ALLEXCEPT()

- ALLSELECTED()

- ALLNOBLANKROW()

Figure 12-8 shows the output from this function applied to countries. It can, of course, be sliced, filtered, and displayed in many different ways.

Ratio of SpareParts and Labor Cost to All Sales by CountryName

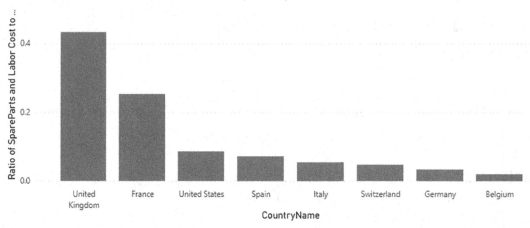

Figure 12-8. Using a moderator function to display a ratio

To continue the principle of trying to make complex function more comprehensible, Figure 12-9 explains how it works in a more visual way.

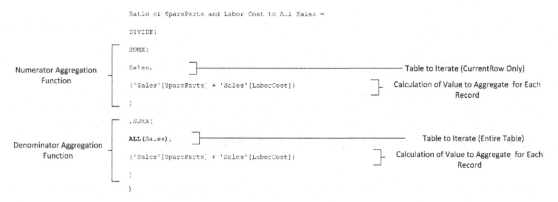

Figure 12-9. Visual explanation of using modifier functions inside an iterator

Count Iterators

To complete the tour of basic iterators, you need to be aware that there are two iterator functions that you can use to count records. These are

- COUNTX() – to count values, dates, or strings

- COUNTAX() – to include logical values in the output

Neither will count blank (empty) values in a field.

Admittedly, these functions are not only simple, but also work in a similar way to their non-iterator forbears COUNT() and COUNTA(). Nonetheless, in the interest of completeness, let's see them applied.

COUNTX() simply needs a table as the first parameter and the field to have non-blank values counted as the second parameter. You can see this in the following short DAX snippet (you can see this in the Client table).

```
Clients With Status Flag = COUNTX('Client', [StatusFlag])
```

You can see this measure applied to a card visual in Figure 12-10.

9
Clients With Status Flag

Figure 12-10. *The COUNTX( ) function*

COUNTAX() also needs a table as the first parameter and the field to have non-blank values counted as the second parameter. You can see this in the following DAX (also in the Client table).

```
Clients With Credit Risk Flag =
COUNTAX('Client', [IsCreditRisk])
```

You can see this measure applied to a card visual in Figure 12-11.

88
Clients With Credit Risk Flag

Figure 12-11. *The COUNTAX( ) function*

If you try and apply COUNTX() instead of COUNTAX() to the IsCreditRisk field, all you will get is an error message.

Available Iterators

There are many more iterator functions (which are also known as "X" functions) in Power BI. Table 12-1 outlines those that are currently available.

It is probably worth taking a look at some more iterators so that you feel at home when using them.

Ranking Output

DAX can do so much when it comes to preparing metrics for BI delivery that it is hard to know exactly what you will need and when. As a more complex example in this short tour of DAX iterators, here is how rank sales by make. I realize that you can do this just by sorting records, but should you need a clear and unequivocal indicator of ranking, then the following DAX snippet (available in the Client table) shows how it can be done:

```
Client Sales Ranking =
RANKX(ALL('Client'), [Total Sales])
```

The output from this function is given in Figure 12-12 where the table is sorted by the Client Sales Ranking field to make the output more intuitive.

Table 12-1. *DAX Iterator Functions*

Formula	Description	Example
MINX( )	Calculates a value for each row in a table and displays the minimum value	MINX(Vehicle, 'Sales'[SpareParts] + 'Sales'[LaborCost])
MAXX( )	Calculates a value for each row in a table and displays the maximum value	MAXX(Vehicle, 'Sales'[SpareParts] + 'Sales'[LaborCost])
SUMX( )	Calculates a value for each row in a table and displays the total	SUMX(Vehicle, 'Sales'[SpareParts] + 'Sales'[LaborCost])
AVERAGEX( )	Calculates a value for each row in a table and displays the average of these values	AVERAGEX(Vehicle, 'Sales'[SpareParts] + 'Sales'[LaborCost])
COUNTX( )	Calculates a value for each row in a table and counts the resulting rows including non-blank results of the calculation	COUNTX(Vehicle, 'Sales'[SpareParts] + 'Sales'[LaborCost])
COUNTAX( )	Calculates a value for each row in a table and counts the resulting rows not including Boolean values	COUNTAX (Vehicle, 'Sales'[SpareParts] + 'Sales'[LaborCost])
GEOMEANX( )	Calculates a value for each row in a table and displays the geometric mean of these values	GEOMEANX(Vehicle, 'Sales'[SpareParts] + 'Sales'[LaborCost])
MEDIANX( )	Calculates a value for each row in a table and displays the median value of these values	MEDIANX(Vehicle, 'Sales'[SpareParts] + 'Sales'[LaborCost])
PERCENTILEX. EXC( )	Returns the percentile of a record relative to the data set	PERCENTILEX.EXC(Vehicle, 'Sales'[SpareParts] + 'Sales'[LaborCost])

(*continued*)

Table 12-1. (*continued*)

Formula	Description	Example
PERCENTILEX. INC()	Returns the percentile of a record relative to the data set	PERCENTILEX.INC(Vehicle, 'Sales'[SpareParts] + 'Sales'[LaborCost])
PRODUCTX()	Returns the product of the numbers in the column	PRODUCTX(Sales, 'Sales'[CostPrice] * RELATED (Geography[RiskFactor]))
RANKX()	Orders the rows by progressive rank	**RANKX(ALL(Vehicle[Make]), SU MX(RELATEDTABLE(InvoiceLin es), [SalePrice]))**
STDEVX.P()	Calculates a value for each row in a table and displays the standard deviation of the entire population of these values	STDEVX.P(Vehicle, 'Sales'[SpareParts] + 'Sales'[LaborCost])
STDEVX.S()	Calculates a value for each row in a table and displays the standard deviation of a sample population of these values	STDEVX.S(Vehicle, 'Sales'[SpareParts] + 'Sales'[LaborCost])
VARX.S()	Calculates a value for each row in a table and displays the variance of the entire population of these values	VARX.S(Vehicle, [SpareParts]+[LaborCost])
VARX.P()	Calculates a value for each row in a table and displays the variance of the entire population of these values	VARX.P(Vehicle, [SpareParts]+[LaborCost])

ClientName	SalePrice	Client Sales Ranking
Laurent Saint Yves	£1,343,950.00	1
Vive La Vitesse	£1,269,600.00	2
Honest Pete Motors	£947,140.00	3
Magic Motors	£602,850.00	4
Glitz	£589,490.00	5
Capots Reluisants S.A.	£583,115.00	6
King Leer Cars	£581,580.00	7
Antonio Maura	£550,330.00	8
Le Luxe en Motion	£572,400.00	8
Alexei Tolstoi	£521,490.00	9
Screamin' Wheels	£540,100.00	9
M. Pierre Dubois	£511,675.00	10
SuperSport S.A.R.L.	£509,630.00	11
La Bagnole de Luxe	£497,190.00	12
Birmingham Executive Prestige Vehicles	£469,740.00	13
Prestige Imports	£456,150.00	14
Andrea Tarbuck	£454,500.00	15
Sondra Horowitz	£428,000.00	16
Matterhorn Motors	£422,515.00	17
Liverpool Executive Prestige Vehicles	£419,650.00	18
Glittering Prize Cars Ltd	£393,980.00	19
Alicia Almodovar	£382,090.00	20

Figure 12-12. Using the RANKX( ) function to classify data from the same table

The RANKX() function needs a little more initial explanation - and then you will also need to understand how other optional parameters can alter the shape of the output.

Essentially - and as for all iterators - you need the two compulsory parameters:

- The table (in this case it is the Client table)

- The value to define the classification or ranking (Total Sales in this example)

As simple as these two elements are, there are some caveats that you need to be aware of at the start.

- Firstly you need, in essence, to know from the start what kind of element you will be ranking the data on. Consequently, if you wish to rank on Make, Model, or color (for instance), you will need to specify the Vehicle table, as these fields are from this particular table. It is important to add as the first parameter the *table that contains the data that you will be using in the output.* You do *not* have to specify which field(s) from this table that you will be using, however.

- Unless the table you are ranking on contains only unique elements in the column that you are using in the output (as is the case in the Client table for instance), you should specify the *column that you will be using in the output* as well as the table name as the first parameter in the RANKX() function.

- You then need to wrap the source table in the ALL() modifier. This is because otherwise the RANKX() function ranks each element (client in this example) compared to *itself* - and *not* compared to all the clients.

- Then you need to decide which numeric value you will be using as the metric for the ranking. If this value is a measure (as is the case here with the Total Sales measure), then the RANKX() function can be as simple as it is in this example.

- If the value is *not a measure,* you need to enter an aggregate function wrapped in CALCULATE(). So whereas the Total Sales measure is written as

```
SUM(Sales[SalePrice])
```

If this measure did not exist, you need to enter it in a RANKX() formula as

```
CALCULATE(SUM(Sales[SalePrice]))
```

All of which gives an alternative formula which could read

```
Client Sales Ranking 2 =
RANKX(ALL('Client'), CALCULATE(SUM(Sales[SalePrice])))
```

Another way of applying the RANKX() function is to use another iterator to deliver the value used for the ranking. The following short piece of DAX gives an example of this (this measure is added to the Vehicle table).

```
SalesRankByMake =
RANKX(ALL(Vehicle[Make]),
SUMX(RELATEDTABLE(Sales), [SalePrice]))
```

Indeed, if you want to simplify the approach, you could use a measure instead of the RELATEDTABLE() function – like this:

```
SalesRankByMake =
RANKX(ALL(Vehicle[Make]),
SUMX([Total Sales], [SalePrice]))
```

Here we are ranking makes by sales. However, the sales figures are not in the same table as the Make field. So you need to navigate through the data model using RELATEDTABLE() to find the value that you want to use to provide the ranking. This DAX snippet reinforces the point that the RANKX() function is an iterator. Consequently if you want to look up values in another table, you have to apply the RELATED() function - assuming that the data model allows the link.

If you apply this measure to a simple table that lists the makes sold and the sale price and then sort by the SalesRankByMake field, you will see something like Figure 12-13. It is worth noting that the table is sorted by the SalesRankByMake field to make the output more instantaneously comprehensible.

Make	SalesRankByMake	TotalSalePrice
Ferrari	1	£5,540,850
Aston Martin	2	£5,102,755
Bugatti	3	£1,915,500
Lamborghini	4	£1,727,150
Bentley	5	£1,689,240
Rolls Royce	6	£1,637,000
Porsche	7	£1,169,240
Jaguar	8	£882,955
Mercedes	9	£471,115
Triumph	10	£396,270
McLaren	11	£295,000
Alfa Romeo	12	£243,510
Lagonda	13	£218,000
Noble	14	£206,900
Delahaye	15	£107,000
Citroen	16	£101,080
Delorean	17	£99,500
Austin	18	£64,500
BMW	19	£60,500
Peugeot	20	£21,045
Morgan	21	£18,500
Trabant	22	£8,440
Reliant	23	£1,900
Maybach	24	
Total	**1**	**£21,977,950**

Figure 12-13. *Using the RANKX( ) function with another iterator*

Ranking Using Multiple Columns

DAX also allows you to rank elements across multiple columns. The approaches are, however, different depending on whether the columns come from the same table or not.

Let's begin with the simplest option which is ranking using multiple columns from the same table. If you take a look at the following DAX and the output that it generates in Figure 12-14, you can see that the overall rank depends on the combination of sales by make and model. This measure is also in the Vehicle table.

```
Make and Model Sales Rank =
RANKX(ALL(Vehicle[Make],Vehicle[Model]), [TotalSalePrice])
```

Make	Model	TotalSalePrice	Make and Model Sales Rank
Bugatti	57C	£1,695,000	1
Ferrari	355	£1,191,450	2
Aston Martin	Virage	£1,078,640	3
Aston Martin	DB6	£1,019,095	4
Ferrari	Enzo	£1,015,000	5
Aston Martin	DB9	£959,500	6
Lamborghini	Diabolo	£905,000	7
Ferrari	Testarossa	£870,000	8
Ferrari	F50	£760,950	9
Bentley	Flying Spur	£678,940	10
Lamborghini	Jarama	£550,000	11
Ferrari	F40	£519,500	12
Porsche	911	£513,750	13
Rolls Royce	Phantom	£481,100	14
Aston Martin	DB2	£456,440	15
Ferrari	360	£363,000	16
Aston Martin	Vanquish	£361,040	17
Rolls Royce	Silver Seraph	£359,450	18
Jaguar	XK120	£346,640	19
Aston Martin	Vantage	£343,500	20

Figure 12-14. *Using the RANKX( ) function over multiple columns from the same table*

If you want to rank data where the ranking columns come from different tables, you will need to extend the approach slightly, using the CROSSJOIN() function. Once again, this is probably best understood with a practical example.

Take a look at the following piece of DAX that you can find in the SalesInfo table.

```
Country and Color Sales Rank =
RANKX(
CROSSJOIN(ALL(Geography[CountryName]), ALL(Vehicle[Color])),
[Total Sales])
```

This delivers the output that you can see in Figure 12-15.

CountryName	Color	Total Sales ▼	Country and Color Sales Rank
United Kingdom	Black	3691590	1
France	Red	1211100	2
France	Black	1198610	3
United Kingdom	Blue	1135930	4
United Kingdom	Night Blue	1014090	5
France	Silver	920325	6
France	British Racing Green	856400	7
United Kingdom	Silver	853280	8
France	Blue	714950	9
United Kingdom	British Racing Green	711075	10

Figure 12-15. Ranking output using columns from separate tables

Here, the measure Country and Color Sales Rank is aggregating the data by Country and Color and then returning the rank of each combination of the two. The only difference is that, to allow RANKX() to use columns from different tables for the ranking, you need to

- Wrap all the column names inside a CROSSJOIN() function as the first parameter of the RANKX() function. This, internally, creates a virtual table of all the possible combinations of country and color. So you are, in effect, supplying a table as the first parameter - just as you have always done when using RANKX(). I realize that this is a very succinct explanation of CROSSJOIN() so rest reassured that I will explain this function in greater detail in Chapter 16.

- Enclose each of the columns inside the CROSSJOIN() function in a separate ALL() function.

Note   The measures that you create using the RANKX( ) function can be placed in any table. Nonetheless, I always prefer to add them to the table that contains the columns used for the output. This seems a more coherent approach. You are, of course, free to add any measure to any table in the dataset using any logic that you wish.

Handling Ties in RANKX( )

Ties in ranked data are situations when two rows have exactly the same value. If you glance ahead at Figure 12-9, you can see that there are multiple models ranked 81.

This table uses the following ranking function (in the SalesInfo table this time):

```
Vehicle Sales Rank =
RANKX(ALL(Vehicle[Model]), [TotalSalePrice])
```

DAX (or more precisely the RANKX() function) gives you a couple of ways of handling this:

- Leave blanks in the numbering scheme (as shown in Figure 12-16).

- Continue the numbering without any breaks.

Model	TotalSalePrice	Model Sales Rank
TR6	£31,440	61
Spider	£48,100	62
E-Type	£39,500	63
TR3A	£39,500	63
Roadster	£36,450	65
250SL	£35,550	66
E30	£33,500	67
Lichfield	£30,100	68
135	£25,500	69
Traaction Avant	£25,000	70
Boxster	£22,500	71
Cambridge	£22,500	71
Alpina	£21,500	73
Plus 4	£18,500	74
1750	£13,525	75
404	£12,945	76
GT6	£12,750	77
175	£12,500	78
Princess	£11,900	79
Rosalie	£10,190	80

Tied Values
Break in Numbering

Figure 12-16. *Ties in the output from the RANKX( ) function*

The default approach in RANKX() is to leave blanks in the numbering scheme and "jump" to the next number after the tied rows in the table. This means that the numbering scheme goes from 71 to 73 as there is no 72, but two records at the number 71 spot.

You can, if you want, override this behavior by adding another parameter to the RANKX() function. You can see this in the following DAX snippet which extends the RANKX() function with two optional parameters.

```
Vehicle Sales Rank Continuous=
RANKX(ALL(Vehicle[Model]), [TotalSalePrice],,ASC,Dense)
```

The two optional parameters are the fourth and fifth that you can add to a RANKX() function (we will gloss over the third parameter for the moment).

- The fourth parameter specifies whether the ranking will be highest to lowest (the default) or lowest to highest. Here you can set the values to be either highest to lowest (DESC) or lowest to highest (ASC).

- The fourth parameter specifies whether the ranking will skip tied rows or not. Here you can set the values to either *Dense* or *Skip* (the default).

As you can see in Figure 12-17, the numbering of the rows in the output is now completely sequential without any breaks once the fifth parameter is set to Dense.

Model	TotalSalePrice	Vehicle Sales Rank Continuous
TR...	£94,440	61
Spider	£48,100	62
E-Type	£39,500	63
TR3A	£39,500	63
Roadster	£36,450	64
250SL	£35,550	65
E30	£33,500	66
Lichfield	£30,100	67
135	£25,500	68
Traaction Avant	£25,000	69
Boxster	£22,500	70
Cambridge	£22,500	70
Alpina	£21,500	71
Plus 4	£18,500	72
1750	£13,525	73
404	£12,945	74
GT6	£12,750	75
175	£12,500	76
Princess	£11,900	77
Rosalie	£10,190	78

Tied Values
No Blank Numbering

Figure 12-17. *Ties in the output from the RANKX( ) function handled by a RANKX( ) parameter*

To end with, take a look at Figure 12-18 that explains the various parameters of the RANKX() function works in a more visual way.

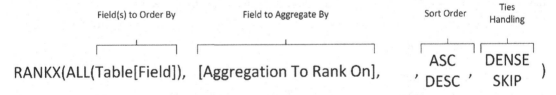

Figure 12-18. *Visual explanation of the RANKX( ) function*

Percentile Calculations Using Iterators

As a final example of an iterator, let's take a look at one of the more advanced statistical iterators that are available in DAX. This is the PERCENTILEX.INC() function.

This function returns the value at a specified percentile threshold in the data. You can see it applied in Figure 12-19. The measure itself is added to the SalesInfo table.

```
25th Percentile =
PERCENTILEX.INC(Sales, 'Sales'[SalePrice], 0.25)
```

CountryName	25th Percentile
Belgium	7,837.50
France	19,950.00
Germany	19,500.00
Italy	12,500.00
Spain	14,450.00
Switzerland	8,587.50
United Kingdom	9,950.00
United States	17,725.00
Total	**12,500.00**

Figure 12-19. *Using the PERCENTILEX.INC() function in DAX*

The output from this measure is shown in Figure 12-19.

While this threshold value is of limited use on its own, it can be used as a "helper" value to deliver more meaningful analyses such as quartiles and deciles. As an example, take a look at the following DAX (added to the SalesInfo table) that calculates sales filtered on amounts less than the 25th quartile. In other words, the following DAX returns sales for the bottom quartile.

```
Lowest Quartile Sales =
CALCULATE([Total Sales],
FILTER('Sales', [SalePrice] < PERCENTILEX.INC(Sales,
'Sales'[SalePrice], 0.25))
)
```

You can see the output as both a card and a table in Figure 12-20.

515K

Lowest Quartile Sales

CountryName	Lowest Quartile Sales
Belgium	11850
France	150315
Germany	29050
Italy	17700
Spain	79880
Switzerland	20725
United Kingdom	196120
United States	88400
Total	**514960**

Figure 12-20. *Applying the PERCENTILEX.INC( ) function to show quartile values*

Conclusion

In this chapter you took a look at DAX iterators. These are another powerful feature of DAX that let you develop custom calculations over rows of data on a row-by-row basis. In many cases they can be used to avoid adding calculated columns to the data model which can make for a lighter and faster model. You then use the measures created using iterators in your visuals to deliver specific insights based on your data.

Firstly you saw how to apply iterator functions so that you can apply a calculation to a set of rows and return an aggregation, be it a sum, average, count, or any other available aggregate function.

You then saw how to filter iterators to produce more tailored outputs. Finally you learned about the RANKX() and PERCENTILEX.INC() functions that enable you to classify data and deliver ordered and grouped output using iterator-based measures.

The functions described in this chapter can be found in the Power BI file PrestigeCarsDimensionalWithIterators.pbix.

It is now time to move on to a core aspect of the data model that you have been using and that was introduced very early in the book. This is the date dimension. The next chapter will not only explain the concept of the date dimension but will also explain ways to create this essential pillar of a functioning data model.

CHAPTER 13

Creating and Applying a Date Dimension

DAX is extremely good at analyzing data over time. This is done using a series of DAX functions that simplify analysis that use dates as a core aspect of the analytics. This kind of analysis is called Time Intelligence.

For time intelligence to work, you will need a *date dimension*. A date dimension is a table that contains an uninterrupted range of dates that begins at least at the *earliest* date, and that ends with a date at least equal to the *final* date in the data that you wish to analyze over time. This table contains not only a range of dates, but columns that isolate parts of every date. These other columns contain elements like the year, month, week, quarter - and, as you will see, many other combinations of the constituent parts of a date. These columns can then be used in visuals not only to display parts of a date but also to aggregate date over time.

In practice a date dimension will nearly always be a table that begins on the 1st of January of the earliest date in your data and that ends on the 31st of December of the last year for which you have data. However this is only a suggestion, not a technical constraint. Once you have built a date table (which is a date dimension now) that spans the complete range of dates that you want to analyze, you can join it to any of the other tables that contain a date field in your data model and then begin to exploit all the time-related analytical functions of Power BI Desktop.

The good news is that you can use Power BI Desktop itself to create a date table. It is also possible to import a contiguous range of dates from other applications such as Excel. As I would prefer you to come to appreciate the breadth and depth of DAX, I will explain how to implement a data table directly in Power BI Desktop.

© Adam Aspin 2023
A. Aspin, *Pro DAX and Data Modeling in Power BI*, https://doi.org/10.1007/978-1-4842-8995-2_13

We will be using the Power BI Desktop file
PrestigeCarsDimensionalNoDateDimension.pbix in this chapter as the basis for adding
and populating a date dimension. This file is one of the sample files that you can
download from the Apress website as outlined in Appendix A.

Why Use a Date Dimension?

Probably the first question to ask, when looking at creating and adding a date dimension
to the data model is: why bother?

There are several very valid reasons for taking this approach - even if it can seem a
little laborious at first sight. Some of these reasons are

- A date dimension enables you to group data by hierarchical time
  elements such as quarters, months, weeks, and days easily.

- The columns of a date dimension can be used in visuals to display
  date parts instead of full dates (this can be useful if you want to use
  abbreviations to save space).

- A date dimension can display custom combinations of date elements
  that you may need in specific visuals.

- A centralized date dimension joined to many tables allows you to
  filter and slice on a date much more easily across the data model.

- Including a date dimension allows the use of DAX time intelligence
  functions. These are built-in (and often extremely simple) DAX
  functions that help you develop measures that analyze data over time
  quickly and easily.

- A date dimension centralizes date values from many tables. This can
  make it easier to track changes over time.

Written down like this, these reasons may seem a little dry. Yet when you see a
date dimension in action later in this chapter and time intelligence applied to solve
real-world challenges in the next chapter, you will hopefully see why creating a date
dimension is worth the little extra effort at the outset.

Note   A date dimension is often referred to as a *time* dimension. Indeed, analytics people talk about analyzing and grouping data over time when they are breaking data down into date elements. So the terms date and time are often used interchangeably in analytics. This is true, of course, until you need both a "date" dimension and a "time" dimension, in which case the "time" dimension is exactly what it is described as – a dimension breaking down the hours, minutes, and seconds in the day.

Creating the Date Table

The first requirement before starting to apply time intelligence is to have a valid date table. The following few steps explain one way to create a date table using - and extending - the DAX skills you acquired in Chapter 7.

1. Open the sample file
   PrestigeCarsDimensionalNoDateDimension.pbix.

2. In Report view click on any table in the View Pane (or switch to Data View), and click the New Table button in the Table tools menu. The expression `Table` = will appear in the Formula Bar.

3. Replace the table name to the left of the equals sign with `Date`.

4. Click to the right of the equals sign and enter the following DAX function

   ```
   Date = CALENDAR("1/1/2018", "31/12/2022")
   ```

5. Press Enter or click the tick icon in the Formula Bar. Power BI Desktop will create a table containing a single column of dates from the 1st of January 2018 until the 31st of December 2022.

6. In the Fields list, right-click on the Date field in the date table and select Rename. Rename the date field to `DateKey`. The date table will look like Figure 13-1.

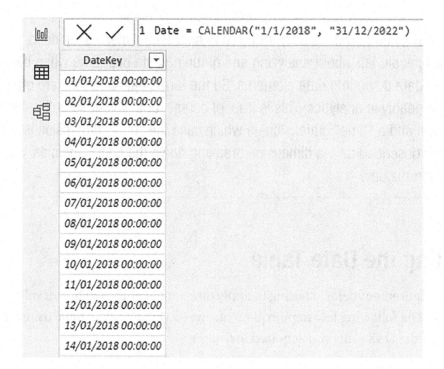

Figure 13-1. *An initial date table for a time dimension*

This initial date dimension used the CALENDAR() function to create a sequential and uninterrupted range of dates. As you saw in the DAX in step 4, the CALENDAR() function takes only two parameters. These are

- *First parameter* - the *start* date

- *Second parameter* - the *end* date

Using these parameters the CALENDAR() function has added a series or rows to a new table. Each row contains a date. The date series starts at the date supplied as the first parameter and continues – uninterrupted - until the date supplied as the second parameter. This gives you a complete range of dates without any blanks in the table to serve as the basis for a date dimension. This initial column of dates is now ready to be extended with other columns to become fully fledged data table.

I have chosen to name this column of dates DateKey. This is only a personal choice (and habit). You can call the main date column whatever you want.

Note   It is *vital* to create a date dimension that encompasses all dates in any
of the tables containing data that you want to analyze over time. If your date
dimension is smaller than the date range of data in your tables of data, you will be
excluding records from your analyses – and falsifying the results.

Marking a Table as a Date Table

The table that you just created may be a table of dates, but it is not yet - at least as far as
Power BI Desktop is concerned - a date table. There is one more simple step to carry out
which is to mark the table containing dates as a date table.

1.   In Table view select the date table in the Fields Pane.

2.   In the Table tools menu, click the Mark as date table button, and
     select Mark as date table in the popup menu. You can see this in
     Figure 13-2.

Figure 13-2. *Setting a table as a date table*

3.   In the Mark as date table dialog, select the date column. You can
     see this in Figure 13-3.

Mark as date table   ✕

Select a column to be used for the date. The column must be of the data type 'date' and must contain only unique values. Learn more

Date column

| DateKey | ⌄ |

✓ Validated successfully

When you mark this as a date table, the built-in date tables that were associated with this table are removed. Visuals or DAX expressions referring to them may break.
Learn how to fix visuals and DAX expressions

OK   Cancel

Figure 13-3. *Setting a table as a date table*

4.   Click OK.

This may be a simple operation to carry out, but it is essential. What you have done allows Power BI to use its considerable range of time intelligence functions to deliver time-based analytics. You will discover these in the next chapter.

Note   It is *fundamental* to set a table of dates as a date table. Without this many of the advanced time intelligence features that you will look at in Chapter 14 will not work correctly unless certain conditions are met. So it is usually best to "play safe" and mark the date dimension as a date table.

Extending the Date Dimension

Now that you have the core of a date table, you can extend it to provide all the ways of looking at a date that you may need either to create visuals or to group and aggregate data over time. This can cover a wide range of requirements, including

- Year elements (showing the year as two or four figures, for instance)

- Quarter elements (showing the number of the quarter with or without decorative text)

- Month elements (showing the month as a number – with or without leading zeroes - or as the month name in full or as an abbreviation)

- Week elements (showing the week as a number – with or without leading zeroes)

- Day elements (showing the day as a number – with or without leading zeroes)

- Date elements (displaying the date in various formats)

- Combination elements (combining year and quarter, year and month, etc.)

These various derived elements should be added as new columns (calculated columns if you prefer to think of them this way) to the date table.

As this collection of DAX expressions can seem a little daunting at first, let's look at them section by section. You can take a look at the finished date table to see the output from each of the functions described below.

Note   The DAX formulas that you see in the following tables are not the only ways of extracting date elements from a date. In some cases there could be several ways of extracting the date attribute. So feel free to use alternatives if you want to. In some cases I "ring the changes" by applying one DAX formula to one element and a different DAX formula to a different element where either could be applied.

In practice, a date dimension could contain dozens - or even hundreds - or ways of breaking down dates. The exact set of date attributes that you will need will, of course, depend on your specific analytical requirements. Consequently the series of data columns (or attributes if you prefer) that are outlined in the next few pages are only an introduction to some of the more common elements that a date dimension may contain. There could be many others that you may choose to add to answer particular time analysis challenges that you may face.

Year Elements

Isolating the year from a date so that you can group data at annual level is mercifully simple in DAX. The two date attributes that are frequently used are defined in Table 13-1. As you can see, the DAX YEAR() function is key, here.

Table 13-1. *Year Attributes in a Date Table*

Column Title	Formula	Comments
FullYear	YEAR([DateKey])	Isolates the year as a four digit number
ShortYear	VALUE(RIGHT(YEAR([DateKey]),2))	Isolates the year as a two digit number

Note   The names of the attribute columns in the date dimension can be anything you want and can contain spaces if this is your preference.

You can add explanatory or decorative text to the year value as well, if you wish.

Quarter Elements

The DAX QUARTER() function allows you to isolate the quarter of the year from a date and then extend if with "decorative" text to display the quarter in various ways for use in visuals. You can see this in Table 13-2.

Table 13-2. *Quarter Attributes in a Date Table*

Column Title	Formula	Comments
QuarterFull	`"Quarter " & QUARTER([DateKey])`	Displays the current quarter
QuarterAbbr	`"Qtr " & QUARTER([DateKey])`	Displays the current quarter as a three letter abbreviation plus the quarter number
Quarter	`"Q" & QUARTER([DateKey])`	Displays the current quarter in short form
QuarterNumber	`QUARTER([DateKey])`	Displays the number of the current quarter. This is essentially used as a sort by column

Month Elements

While the DAX MONTH() function returns the number of the month in the year, it is the FORMAT() function that returns the month as a text. So you will need to use the appropriate function depending on the output that you require. You can see this in Table 13-3.

Table 13-3. *Month Attributes in a Date Table*

Column Title	Formula	Comments
MonthNumber	`MONTH([DateKey])`	Isolates the number of the month in the year as one or two digits
MonthNumberFull	`Format([DateKey], "MM")`	Isolates the number of the month in the year as two digits with a leading zero for the first nine months
MonthFull	`Format([DateKey], "MMMM")`	Displays the full name of the month
MonthAbbr	`Format([DateKey], "MMM")`	Displays the name of the month as a three letter abbreviation

Week Elements

DAX can also isolate the week of the year to allow you to aggregate data at a week level. This is described in Table 13-4. The main DAX function used here is WEEKNUM().

Table 13-4. *Week Attributes in a Date Table*

Column Title	Formula	Comments
WeekNumber	WEEKNUM([DateKey])	Shows the number of the week in the year
WeekNumberFull	FORMAT(WEEKNUM([DateKey]), "00")	Shows the number of the week in the year with a leading zero for the first nine weeks

Day Elements

The DAY() function is the key to extracting the day from a date as you can see in Table 13-5. However, you will need the WEEKDAY() and FORMAT() functions as well in some cases.

Table 13-5. *Day Attributes in a Date Table*

Column Title	Formula	Comments
DayOfMonth	DAY([DateKey])	Displays the number of the day of the month
DayOfMonthFull	FORMAT(DAY([DateKey]),"00")	Displays the number of the day of the month with a leading zero for the first nine days
DayOfWeek	WEEKDAY([DateKey])	Displays the number of the day of the week
DayOfWeekFull	FORMAT([DateKey],"dddd")	Displays the name of the weekday
DayOfWeekAbbr	FORMAT([DateKey],"ddd")	Displays the name of the weekday as a three letter abbreviation

There is one peculiarity of the WEEKDAY() function that you need to be aware of. This is that, by default, it is set to start with Sunday as day 1 (Monday as day 2, etc.)

This can be overridden through adding a second parameter, like this:

- WEEKDAY([DateKey]. 1) gives the same result as
  WEEKDAY([DateKey]) where Sunday is 1, Monday is 2 until
  Saturday as 7.

- WEEKDAY([DateKey], 2) sets Monday as 1, Tuesday as 2 until
  Sunday as 7.

- WEEKDAY([DateKey], 3) Sets Sunday as 0, Monday as 1 until
  Sunday as 6.

Date Elements

Even simple dates may need to be displayed in different ways. Table 13-6 gives a couple of examples of this.

Table 13-6. *Date Attributes in a Date Table*

Column Title	Formula	Comments
ISODate	[FullYear] & [MonthNumberFull] & [DayOfMonthFull]	Displays the date in the ISO (internationally recognized) format of YYYYMMDD
FullDate	[DayOfMonth] & " " & [MonthFull] & " " & [FullYear]	Displays the full date with spaces

A full date can use any separators that you want. You can even create multiple full date columns (with slightly different names, of course) to display a full date in different ways.

Combination Elements

Here, as we are back in the simple world of calculated columns, combining date elements is as simple as creating a new column that refers to columns that you have already added to the date table. You can see some date attributes combined in this way in Table 13-7.

315

Table 13-7. *Combination Attributes in a Date Table*

Column Title	Formula	Comments
QuarterAndYear	[Quarter] & " " & [FullYear]	Shows the quarter and the year
QuarterAndYearAbbr	[Quarter] & " " & [ShortYear]	Shows the quarter and the year with the year as two figures
MonthAbbrAndYearAbbr	[MonthAbbr] & " " & [ShortYear]	Shows the abbreviated month and year
MonthAbbrAndYear	[MonthAbbr] & " " & [FullYear]	Shows the abbreviated month and year as four figures
MonthAndDay	[MonthAbbr] & " " & [DayOfMonth]	Displays the abbreviated month and the day of the month
MonthAndDayNumber	[MonthNumberFull] & [DayOfMonthFull]	Shows the month and the day of the month numbers. This is essentially used as a sort by column
YearAndQuarterNumber	[FullYear] & [QuarterNumber]	Shows the year and quarter numbers. This is essentially used as a sort by column
YearAndWeek	VALUE([FullYear] &[WeekNumberFull])	Indicates the year and week. The VALUE() function ensures that the figure is considered as numeric by Power BI Desktop
YearAndMonthNumber	VALUE([FullYear] & [MonthNumberFull])	A numeric value for the year and month. The VALUE() function ensures that the figure is considered as numeric by Power BI Desktop

Add the new columns containing the formulas explained in Tables 13-1 to 13-7. As Chapter 7 provided an exhaustive explanation covering the techniques that you need to add new columns, I will not repeat the process in detail here.

The first few columns of the date table should now look like Figure 13-4.

DateKey	FullYear	ShortYear	QuarterNumber	Quarter	QuarterFull	QuarterAbbr	MonthNumber	MonthNumberFull	MonthFull	MonthAbbr	WeekNumber	WeekNumberFull	DayOfMonth	DayOfMonthFull	DayOfWeek	DayOfWeekFull
01/01/2018 00:00:00	2018	18	1	Q1	Quarter 1	Qtr 1	1	01	January	01	1	01	1	01	2	Monday
02/01/2018 00:00:00	2018	18	1	Q1	Quarter 1	Qtr 1	1	01	January	01	1	01	2	02	3	Tuesday
03/01/2018 00:00:00	2018	18	1	Q1	Quarter 1	Qtr 1	1	01	January	01	1	01	3	03	4	Wednesday
04/01/2018 00:00:00	2018	18	1	Q1	Quarter 1	Qtr 1	1	01	January	01	1	01	4	04	5	Thursday
05/01/2018 00:00:00	2018	18	1	Q1	Quarter 1	Qtr 1	1	01	January	01	1	01	5	05	6	Friday
06/01/2018 00:00:00	2018	18	1	Q1	Quarter 1	Qtr 1	1	01	January	01	1	01	6	06	7	Saturday
07/01/2018 00:00:00	2018	18	1	Q1	Quarter 1	Qtr 1	1	01	January	01	2	02	7	07	1	Sunday
08/01/2018 00:00:00	2018	18	1	Q1	Quarter 1	Qtr 1	1	01	January	01	2	02	8	08	2	Monday
09/01/2018 00:00:00	2018	18	1	Q1	Quarter 1	Qtr 1	1	01	January	01	2	02	9	09	3	Tuesday
10/01/2018 00:00:00	2018	18	1	Q1	Quarter 1	Qtr 1	1	01	January	01	2	02	10	10	4	Wednesday
11/01/2018 00:00:00	2018	18	1	Q1	Quarter 1	Qtr 1	1	01	January	01	2	02	11	11	5	Thursday
12/01/2018 00:00:00	2018	18	1	Q1	Quarter 1	Qtr 1	1	01	January	01	2	02	12	12	6	Friday
13/01/2018 00:00:00	2018	18	1	Q1	Quarter 1	Qtr 1	1	01	January	01	2	02	13	13	7	Saturday
14/01/2018 00:00:00	2018	18	1	Q1	Quarter 1	Qtr 1	1	01	January	01	3	03	14	14	1	Sunday
15/01/2018 00:00:00	2018	18	1	Q1	Quarter 1	Qtr 1	1	01	January	01	3	03	15	15	2	Monday
16/01/2018 00:00:00	2018	18	1	Q1	Quarter 1	Qtr 1	1	01	January	01	3	03	16	16	3	Tuesday
17/01/2018 00:00:00	2018	18	1	Q1	Quarter 1	Qtr 1	1	01	January	01	3	03	17	17	4	Wednesday
18/01/2018 00:00:00	2018	18	1	Q1	Quarter 1	Qtr 1	1	01	January	01	3	03	18	18	5	Thursday
19/01/2018 00:00:00	2018	18	1	Q1	Quarter 1	Qtr 1	1	01	January	01	3	03	19	19	6	Friday
20/01/2018 00:00:00	2018	18	1	Q1	Quarter 1	Qtr 1	1	01	January	01	3	03	20	20	7	Saturday
21/01/2018 00:00:00	2018	18	1	Q1	Quarter 1	Qtr 1	1	01	January	01	4	04	21	21	1	Sunday
22/01/2018 00:00:00	2018	18	1	Q1	Quarter 1	Qtr 1	1	01	January	01	4	04	22	22	2	Monday
23/01/2018 00:00:00	2018	18	1	Q1	Quarter 1	Qtr 1	1	01	January	01	4	04	23	23	3	Tuesday
24/01/2018 00:00:00	2018	18	1	Q1	Quarter 1	Qtr 1	1	01	January	01	4	04	24	24	4	Wednesday

Figure 13-4. *The completed date table*

Just to reinforce the point - you have created all these ways of expressing dates and parts of dates so that you can now use them in your tables, charts, and gauges to aggregate and display data over time. Any record that has a date element can now be expressed visually not just as the date itself, but aggregated as years, quarters, months, or weeks. The trick is to prepare all the time groupings that you are likely to need in the date table ready for your dashboards. However you do not need to worry if you find yourself needing an extra column or two a little later as you can always add other columns that contain further date elements later.

Adding Sort by Columns to the Date Table

In some cases you may need to sort a column using the data in another column to provide the sort order. This technique is essential when dealing with date tables, as you want to be sure that any visualizations that contain date elements appear in the right order. The classic example is months. As things stand, if you were to use the MonthFull or MonthAbbr columns in a chart or table, then you would see the month names appearing on an axis or in a column in *alphabetical order*. This is extremely disconcerting and needs to be corrected if you want to keep users (and bosses) onside.

The solution is to set a Sort by column for any column in the date table that uses a "wrong" (which usually means unexpected and visually disconcerting) sort order. Here is an example of what to do.

1. In the Power BI Desktop file where you created the date dimension (or if you have not done this open the sample file PrestigeCarsDimensionalWithDateDimension.pbix).

2. Switch to Data view (unless you are still in Data view).

3. Click on the column title for the QuarterAndYear column.

4. In the Column tools ribbon, click the Sort by column button.

5. Select YearAndQuarterNumber from the popup list.

Table 13-8 gives you the required information to extend the data table with Sort by column settings so that all date elements are sorted correctly. Of course, if you have added your own custom date columns, you will need to add any required Sort by columns.

Table 13-8. The Sort By Columns Needed for the Date Table

Column	Sort By Column
MonthFull	MonthNumber
MonthAbbr	MonthNumber
DayOfWeekFull	DayOfWeek
DayOfWeekAbbr	DayOfWeek
Quarter And Year	YearAndQuarterNumber
FullDate	DateKey
MonthAbbrAndYearAbbr	YearAndMonthNumber
MonthAbbrAndYear	YearAndMonthNumber
MonthAndDay	MonthAndDayNumber

Note Applying Sort by columns implies that you have added them to the date table in the first place. They are not added automatically, and so you will have to add any Sort by columns that are required.

Adding the Date Table to the Data Model

Now that you have a date table, you can integrate it with your data model so that you can start to apply some of the time intelligence that DAX makes possible.

1. Click on the Model View icon to display the tables in the data model.

2. In the Home ribbon click the Manage Relationships button. The Manage Relationships dialog will appear.

3. Click the New button. The Create Relationship dialog will appear.

4. At the top of the dialog, select the Date table from the popup list.

5. Once the sample data from the Date table is displayed, click inside the ISODate column to select it.

6. Under the Date table sample data, select the Sales table from the popup list.

7. Once the sample data from the Sales table is displayed, click inside the Date_SK column to select it. The dialog will look like Figure 13-5.

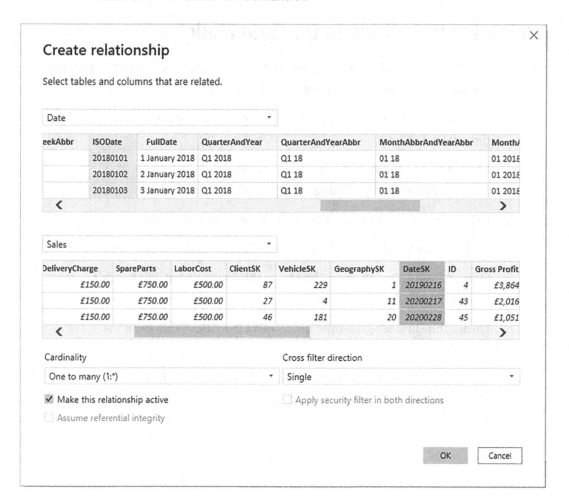

Figure 13-5. *Adding a date dimension to the data model*

8. Click OK. The relationship will appear in the Manage Relationships dialog.

9. Click Close. The date table will appear in the data model joined to the Sales table.

Once the date dimension is a coherent part of the data model, it can be used to group and filter data and, above all, it allows the use of the DAX time intelligence functions that you will discover in the following chapter.

Note   One thing to note is that you do *not* need to join the date dimension to the data model in a date or datetime field. So the join field can be a surrogate key (as is the case in the Prestige Cars Dimensional data model) or is could be through joining the DateKey field to a date field in another table.

Using a Date Table

Now that you have gone to the trouble of creating a date table, and before discovering how to use it to create time-based analytics using DAX time intelligence - let's see succinctly how a data table can help you to create visualizations.

1. In the Power BI Desktop file where you created the date dimension (or if you have not done this open the sample file PrestigeCarsDimensionalWithDateDimension.pbix).

2. Add an empty stacked column chart visual.

3. From the Sales table add the SalePrice field.

4. From the Date table add the QuarterAndYear field.

5. Click on the ellipses at the top right of the chart and then Sort axis => QuarterAndYear.

6. Click on the ellipses at the top right of the chart and then Sort axis => Sort ascending.

The chart should look like the one shown in Figure 13-6.

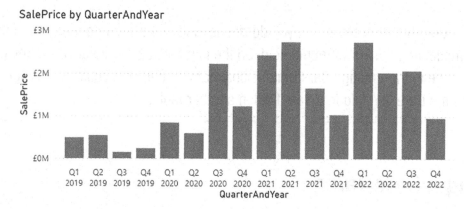

Figure 13-6. *Using attributes from a date table in visuals*

As you can see, the date dimensions have been used to aggregate data by a date element from the date table. So you can now use any field from the date table either to aggregate data at a higher level of granularity than the actual date or to display the date in a different format.

Alternative Date Table Generation Techniques

When using date tables to invoke time intelligence in DAX, there are two fundamental principles that must always be applied. I realize that I mention them elsewhere, but they are so essential that they bear repetition.

- The date range must be *continuous* - that is there must not be any dates missing in the column that contains the list of calendar days in the table of dates.

- The date range must encompass *all the dates* that you will be using in time intelligence functions in other tables in the data model.

The first requirement is covered by the use of the CALENDAR() function to create a date table. The second can require that you discover the lower and upper date thresholds in one or more tables of dates. As this can be a little laborious (not to mention error-prone), it is probably easier to get DAX to find these dates for you and apply them to the CALENDAR() function.

Consequently, once you know where the lowest and highest dates are in your data model (even if they are in separate tables), you can use these dates as the lower and upper boundaries of the CALENDAR() function. In the case of the Prestige Cars data, the formula could read

```
Date =
CALENDAR(
MIN('Stock'[PurchaseDate]),  MAX('Invoices'[InvoiceDate])
)
```

This formula shows that you can apply two functions that you have seen already - MIN() and MAX()- with date and datetime data types. They simply tell DAX to find the earliest and latest dates in a column.

Alternatively, and as a nod to best practice, you may prefer to ensure that the date dimension always covers entire years of dates. In this case the formula to create the table could be

```
Date =
CALENDAR(
STARTOFYEAR('SalesInfo'[DateBought]), ENDOFYEAR('SalesInfo'[DateBought])
)
```

This formula extends the DAX calculation using two functions that were briefly introduced in Chapter 7. These are

- STARTOFYEAR()- This function deduces the first day of the year.

- ENDOFYEAR()- This function deduces the last day of the year.

This formula also shows you that you can define a date range using different columns, or even different tables when you are specifying a date range. This is particularly useful when defining a date table.

Note   The principal advantage of looking up the date boundaries like this is that they will update automatically as the source data changes. So if the Sales table is extended in the data source to contain further rows with dates outside the existing date range, then the Date table will also grow to encompass these new dates.

Alternatively you can use the CALENDARAUTO() function to create a date table. This will, however, scan every table for dates and use the smallest and the largest dates that it finds as the basis for the range of dates in the date dimension. This may include dates that you do not need Time Intelligence for - and consequently create a data table that is too large.

Creating a data table can be fun, but nonetheless takes a few minutes. So one tip I can give you is that you create a Power BI Desktop file that contains nothing but a date dimension table (using a manually defined start and end date just as you saw at the beginning of this example) with all the other columns added. You can then make copies of this "template" file and use them as the basis for any new data models that you create. This can include replacing the fixed threshold dates with references to the data in tables as mentioned previously.

Importing a Date Dimension

You can, if you prefer, use an imported table of dates (and date attributes) as a date dimension. However, you need to be aware of a couple of major caveats:

- Any imported table must have a date data type column containing a range of dates.

- This column of dates must *not* have any dates missing - that is, it must be a complete range of dates.

- Any date attribute columns (i.e., all the other columns containing date parts) must be the correct data type.

If these conditions are met, then you can import a date table from Excel, a CSV file, or a relational database just as you would any other data table or define the date table in Power Query.

Once the external date table has been imported, you will still have to

- Set it as a date table

- Define any Sort by columns

- Join it to other tables in the data model - as explained in the following section.

Conclusion

This chapter introduced you to a core aspect of the Power BI data model - the Date dimension. This artifact is fundamental to implementing data analysis over time.

Firstly you saw how to prepare the dataset for time intelligence by adding a date table. Then you saw how to extract date and time elements from the main date columns. The new columns that you created are then used to filter or group data in your visualizations. This way you can provide a daily, weekly, monthly, quarterly, and yearly breakdown of your source data. These date elements can also help you to filter the datasets that you are using by dates, date elements, and date ranges.

Finally you saw how to join this table to the other tables in the data model. This gave you a complete and coherent data model ready for time analytics. This is the subject of the next chapter.

If you wish to look in detail at the date dimension that was described in this chapter (or use it as a templatemodel), it is available in the downloads for this book. It is called PrestigeCarsDimensionalWithDateDimension.pbix.

CHAPTER 14

Time Intelligence

Much data analysis - and nearly all business intelligence - involves looking at how metrics evolve over time. You may need to aggregate sales by month, week, or year for instance. Perhaps you want to compare figures for a previous month, quarter, or year with the figures for a current period. Whatever the exact requirement, handling time (by which we nearly always mean dates) is essential in Power BI Desktop.

Initially using time functions in Power BI Desktop may only be limited to extracting time intervals from the available data and grouping results by units of time such as days, weeks, months, quarters, and years. As you will find out in the first part of this chapter, DAX makes this kind of analysis really simple.

However Power BI Desktop can also add what is called "Time Intelligence" to data models. This massively useful capability can take your analysis to a fundamentally higher level. This approach necessitates adding a date dimension to the data model (which you learned to do in Chapter 13) and then adding DAX measures based on a specific family of functions that enable you to see how data evolves over time. This chapter is an introduction to time intelligence using a wide range of DAX formulas that are available to help you create time-based calculations quickly and easily.

In this chapter we will develop the Power BI Desktop file PrestigeCarsDimensionalWithDateDimension.pbix file that contains all the columns and measures created so far in this book as well as a complete date dimension. This file is available on the Apress website as described in Appendix A.

Adding Time Intelligence to a Data Model

I want now to explain the vital set of functions that concern time, or rather, the dates used in analyzing data. Power BI Desktop calls this time intelligence (even though it nearly always refers to is the use of date ranges). Applying this kind of temporal analysis

© Adam Aspin 2023
A. Aspin, *Pro DAX and Data Modeling in Power BI*, https://doi.org/10.1007/978-1-4842-8995-2_14

can be a fundamental aspect of data presentation in business intelligence. After all, what enterprise does not need to know how this year's figures compare to last year's and what kind of progress is being made?

Time intelligence will always require a valid date table, which is one of the reasons why we spent a certain amount of time describing this core pillar of a successful data model in the previous chapter. The good news is that once you have a valid date table that is integrated into the data model, and have acquainted yourself with a handful of date and time functions in DAX, you can very quickly deliver some extremely impressive analytical output. These kinds of calculations can cover (among other things)

- YearToDate, QuarterToDate, and MonthToDate calculations

- Comparisons with previous years, quarters, or months

- Rolling aggregations over a period of time, such as the sum for the last three months

- Comparison with a parallel period in time, such as the same month in the previous year

- Specifying data ranges for data comparisons

- And many other comparisons for data across time periods

An introduction to time intelligence in Power BI Desktop will give you a taste of some of the DAX functions that you are likely to use when analyzing data over time. To begin with you will learn how to use the simpler time intelligence functions to deliver time-based analytics in a few seconds. Then you will take a look at some of the more advanced calculations that you can create to analyze data over time.

Applying Time Intelligence

Now that a date dimension is set up, it is (finally) time to see just how DAX can make your life easier when it comes to calculating metrics over time. This is possible only because you have a date table in place and it is connected through the data model to the requisite date field in the table (or tables) that contain the data you want to aggregate. However with the foundations in place, you can now start to deliver some really interesting and persuasive output.

YearToDate, QuarterToDate, and MonthToDate Calculations

To begin with, let's resolve a simple but frequent requirement - calculating month to date, quarter to date, and year to date sales figures.

The three functions that you will see in this example are extremely similar. Consequently I will only explain the first one (a month to date calculation), and then will let you create the next two in a couple of copy, paste, and tweak operations.

1. Open the file PrestigeCarsDimensionalWithDateDimension.pbix.

2. In the Fields Pane, select the Sales table.

3. Create a new measure containing the following DAX:

   ```
   MonthSales = TOTALMTD(SUM(Sales[SalePrice]),'Date'[DateKey])
   ```

4. Press Enter or click the tick mark icon to complete the measure definition.

Now copy the formula that you just created and use it as the basis for two new measures that you can also add to the Sales table. These will be quarterly sales to date and annual sales to date. The formulas are

```
QuarterSales = TOTALQTD(SUM(Sales[SalePrice]), 'Date'[DateKey])
YearSales = TOTALYTD(SUM(Sales[SalePrice]), 'Date'[DateKey])
```

The three formulas that you have used are TOTALMTD() for the Month to Date aggregation, TOTALQTD() for the Quarter to Date aggregation, and TOTALYTD() for the Year to Date aggregation. All three functions take two parameters:

- The *aggregate function* to use (we are using SUM() here — although it could be AVERAGE() or COUNT() or any other of the aggregate functions that you saw in Chapter 7), depending on the actual metric that you want to deliver and the table and column that is aggregated.

- The *key field of the date table*. Since the Sales table is linked to the date table in the data model (using the DateKey field in this data model), DAX can apply the correct calculation if - and only if - you have added a date table to the data model and then *specified the key field of the date table as the second parameter* of the time intelligence function.

Note   Time intelligence functions always need to know which is the date table and which field in this table of uninterrupted unique dates is the date key. So you will always be adding a reference to this column inside the time intelligence function. In the current data model this field is 'Date'[DateKey]. This simple pointer is all the DAX needs to enable time intelligence. Yet, as simple as it may be, it is nonetheless fundamental, and you *must* indicate the correct field when building your own time intelligence functions. However, the date table key field does *not* have to be the field that is used to join the date table to other tables in the data model.

As you have created these measures, I imagine that you would like to see them in action. Figure 14-1 shows the Quarter to date and Year to date sales for 2020 (along with the aggregated sales from the initial SalePrice column in the Sales table) in a simple Power BI Desktop. This output is filtered for 2020 only.

MonthFull	MonthSales	QuarterSales	YearSales
January	451150	451150	451150
February	171040	622190	622190
March	220500	842690	842690
April	258750	258750	1101440
May	99030	357780	1200470
June	237300	595080	1437770
July	608240	608240	2046010
August	1051130	1659370	3097140
September	571065	2230435	3668205
October	316700	316700	3984905
November	111890	428590	4096795
December	808490	1237080	4905285

Figure 14-1. The month, quarter, and year to date functions in DAX

As you can see in this figure, you have each month's sales figures along with the cumulative sales for each quarter to date (and restarting each quarter). The final column shows you the yearly sales total for each month to date. Also you will note that the months appear in calendar order as the MonthFull field has used the MonthNumber field as its Sort By column.

Conceptually the three functions that are outlined above -TOTALMTD(), TOTALQTD(), and TOTALYTD() - can be considered as CALCULATE() functions that have been extended to deliver a specific result for a time frame. You can always calculate aggregations for a period to date using the CALCULATE() function if you want to. However as this "shorthand" version is so easy to use and so practical, I see no reason to try anything more complicated unless there is a real need.

Note   In this example I suggest starting the process of adding a new measure with the Sales table selected merely because this table seems a good place to store the metric. You can create the metric in virtually any table in practice.

Fiscal Year Cumulative Calculations

The TOTALYTD() function is blissfully simple to apply. Yet it has even more capabilities than you may suspect at first sight. This function can also calculate fiscal year cumulative totals (or financial year totals if you prefer) extremely easily. By fiscal or financial years DAX means that you do not want to take the first of January as the year start date and the 31st of December as the year end data but want to specify another 12-month period.

To set a year end other than the default 31st of December for the cumulative aggregation, all you have to do is to add a third parameter to the function. This third optional parameter is a text string that contains the day and month that sets the end of the financial year. So you could use DAX like this (in the Sales table, for instance):

```
Financial Year Sales =
TOTALYTD(SUM(Sales[SalePrice]),'Date'[DateKey], "30/09")
```

You can see the output in Figure 14-2.

MonthFull	SalePrice	Financial Year Sales
January	£451,150.00	689790
February	£171,040.00	860830
March	£220,500.00	1081330
April	£258,750.00	1340080
May	£99,030.00	1439110
June	£237,300.00	1676410
July	£608,240.00	2284650
August	£1,051,130.00	3335780
September	£571,065.00	3906845
October	£316,700.00	316700
November	£111,890.00	428590
December	£808,490.00	1237080

Beginning of New Financial Year

Figure 14-2. Adding a financial year ens date to the TOTALYTD() function

---

Note   Any day and month combination that Power BI can recognize can be used as the third parameter.

Comparisons with Previous Time Periods

While the various DAX functions that return a "previous" timespan are extremely useful, it could be argued that they are a little rigid for some types of calculation. After all comparing with the previous month nearly always requires you to display data by month. So DAX has other alternative methods of comparing data over time that do not require you to specify exactly which time element (day, month, quarter, or year) you want to compare with. Instead you can merely say that you want to go back a defined period in time - and then depending on the choice of time element that you use in a visualization, DAX will automatically calculate the correct figure for comparison.

Looking at Data for the Same Time Point During the Previous Year

Year on year figures can be compared at many levels in a time hierarchy. You could want (or be asked) to compare metrics at day, quarter, month, or week level. Creating different measures for all these time elements would be a little laborious. So DAX has a "one size fits all" function that can be extremely useful. This is SAMEPERIODLASTYEAR().

As is the case for so many other DAX functions, SAMEPERIODLASTYEAR() is best appreciated through seeing a practical example. So try creating the following measure. I suggest that you do this in the Sales table (still using the Power BI Desktop file PrestigeCarsDimensionalWithDateDimension.pbix and still adding the measure to the Sales table).

```
LastYearSales =
CALCULATE(SUM(Sales[SalePrice]), SAMEPERIODLASTYEAR('Date'[DateKey]))
```

This measure uses CALCULATE() to return the total sales for the previous year, filtered to display whichever period the visual is displaying. All that is required to get the SAMEPERIODLASTYEAR() function to work is to indicate the date key field as the parameter. This way the SAMEPERIODLASTYEAR() function filters the output inside the CALCULATE() function and displays the time-shifted output.

This measure can then be used at any level of a date hierarchy. Figure 14-3 shows you the same measure used when creating tables of sales by quarter, month, week, and day. This data is filtered to show data for 2020.

QuarterAbbr	SalePrice	LastYearSales
Qtr 1	£842,690.00	£490,950
Qtr 2	£595,080.00	£541,500
Qtr 3	£2,230,435.00	£149,885
Qtr 4	£1,237,080.00	£238,640
Total	**£4,905,285.00**	**£1,420,975**

MonthAbbrAndYear	SalePrice	LastYearSales
Jan 2020	£451,150.00	£304,950
Feb 2020	£171,040.00	£31,000
Mar 2020	£220,500.00	£155,000
Apr 2020	£258,750.00	£145,100
May 2020	£99,030.00	£373,450
Jun 2020	£237,300.00	£22,950
Jul 2020	£608,240.00	£107,185
Aug 2020	£1,051,130.00	£5,500
Sep 2020	£571,065.00	£37,200
Oct 2020	£316,700.00	£146,190
Nov 2020	£111,890.00	£22,950
Dec 2020	£808,490.00	£69,500
Total	**£4,905,285.00**	**£1,420,975**

WeekNumber	SalePrice	LastYearSales
1	£349,500.00	£84,950
2	£44,700.00	
4	£56,950.00	£220,000
6		£19,500
8	£167,390.00	£11,500
9	£3,650.00	
11		£29,500
12	£220,500.00	
13		£49,500
14		£112,500
15	£102,950.00	£19,600
18	£155,800.00	£89,000
20		£169,500
21		£8,950
22	£99,030.00	£195,000

MonthAbbr	DayOfMonth	SalePrice	LastYearSales
Jan	1	£349,500.00	
Jan	2		£84,950
Jan	7	£32,050.00	
Jan	9	£12,650.00	
Jan	22	£56,950.00	
Jan	25		£220,000
Feb	3		£19,500
Feb	16	£39,500.00	£11,500
Feb	17	£71,890.00	
Feb	22	£56,000.00	
Feb	28	£3,650.00	
Mar	14		£29,500
Mar	20	£220,500.00	
Mar	24		£49,500
Mar	30		£76,000

Figure 14-3.   Displaying previous year figures using SAMEPERIODLASTYEAR()

It is worth noting that you can use any field from the date table in visuals that use time intelligence functions as the Sales table is linked to the date table in the sample data model. To labor the point, once again it was essential to create a coherent and complete data model in order to make time intelligence work perfectly.

Tip   If you want to check that these measures are accurate, simply remember one of the results for the previous year sales and click 2019 in the slicer. You will display the figures for the previous year and can check that the calculation is working correctly.

The measure that you created here is, in fact, a CALCULATE() function that is very similar to the CALCULATE()-based measures that you created in previous chapters. CALCULATE() is simply returning an aggregate. However the SAMEPERIODLASTYEAR() function is using the date dimension to provide a range of dates as a filter. This range of dates is the "jump back in time" to the date period that is displayed in the visual for the previous year. Essentially, time intelligence functions save you having to work out complex ranges of dates to filter on based on the date element used in a visual. Once again, the actual time intelligence function needs you to use the key column of the data dimension as its second parameter.

Comparing with the Same Date Period from a Different Quarter, Month, or Year

In the last section you have seen how to compare data with data from the prior year, quarter, month, or day. The set of time intelligence functions that you just saw are an easy solution if you want to compare with the year, quarter, month, or day immediately before the current year, quarter, month, or day. So what can you do if you want to skip back over two or more time periods?

DAX has another method for these kinds of calculations that can be both easy to implement and extremely powerful. Moreover it can serve as the basis for comparison with either Years, Quarters, or Months. This is the PARALLELPERIOD() function.

As this is a function that has greater potential (and, consequently, more potential for variations) than SAMEPERIODLASTYEAR(), I prefer to give you three examples of measures that you can add to the Sales table at once. These are

Sales for the previous month but one:

```
SalesPrevMthButOne = CALCULATE(SUM(Sales[SalePrice]), PARALLELPERIOD('Date'
[DateKey],-2,MONTH))
```

Sales for four quarters ago:

```
Sales4QtrsAgo = CALCULATE(SUM(Sales[SalePrice]), PARALLELPERIOD('Date'[Date
Key],-4,QUARTER))
```

Sales for three years ago:

```
Sales3YearsAgo = CALCULATE(SUM(Sales[SalePrice]),
PARALLELPERIOD('Date'[DateKey],-3,YEAR))
```

You can see the first two of these functions displayed as simple visualizations in Figure 14-4. All are added to the Sales table in the sample files.

MonthAbbrAndYear	SalePrice	SalesPrevMthButOne	QuarterSales	Sales4QtrsAgo
Jan 2019	£304,950.00		304950	
Feb 2019	£31,000.00		335950	
Mar 2019	£155,000.00	304950	490950	
Apr 2019	£145,100.00	31000	145100	
May 2019	£373,450.00	155000	518550	
Jun 2019	£22,950.00	145100	541500	
Jul 2019	£107,185.00	373450	107185	
Aug 2019	£5,500.00	22950	112685	
Sep 2019	£37,200.00	107185	149885	
Oct 2019	£146,190.00	5500	146190	
Nov 2019	£22,950.00	37200	169140	
Dec 2019	£69,500.00	146190	238640	
Jan 2020	£451,150.00	22950	451150	490950
Feb 2020	£171,040.00	69500	622190	490950
Mar 2020	£220,500.00	451150	842690	490950
Apr 2020	£258,750.00	171040	258750	541500
May 2020	£99,030.00	220500	357780	541500
Jun 2020	£237,300.00	258750	595080	541500
Jul 2020	£608,240.00	99030	608240	149885
Aug 2020	£1,051,130.00	237300	1659370	149885
Sep 2020	£571,065.00	608240	2230435	149885
Oct 2020	£316,700.00	1051130	316700	238640
Nov 2020	£111,890.00	571065	428590	238640
Dec 2020	£808,490.00	316700	1237080	238640
Jan 2021	£569,450.00	111890	569450	842690
Feb 2021	£802,750.00	808490	1372200	842690

Sales for 2 Months Previous

Cumulative Over Current Quarter

Quarterly Sales from 4 Quarters Ago

Figure 14-4. Using the PARALLELPERIOD() function

PARALLELPERIOD() - like SAMEPERIODLASTYEAR() - will adjust to the date level that you have selected in the output. So you can use it at day, month, quarter, or year level in your output. However, you need to be aware that the PARALLELPERIOD() function will compare the current data not just up to the same date in the previous time period (be this a year, a quarter, or a month), but for the *whole* previous time period.

So what PARALLELPERIOD() is doing is to jump back in time the specified number and type of time periods and then returns the selected aggregation up until the *end* of the chosen time period. This opens up multiple possibilities as you can compare different time periods both forwards and backwards in time.

Should you wish to compare with a future time period (compared to the date displayed in a visual), simply ensure that the second parameter of the PARALLELPERIOD() function is set to a positive number.

Comparing a Metric with the Result from the Previous Time Period

Let's push time-based data analysis a little further and imagine that you need to look at how figures have evolved compared to a time interval that immediately precedes a current time period. By this I mean that you want to see how sales for the current month have fluctuated compared to, say, the previous month. All of this can be done quickly and easily using more of the DAX time intelligence functions.

To be precise, there are a set of time intelligence functions that allow you to look back at the time period before the current one. These functions cover analytics by

- Year
- Quarter
- Month
- Day

As is always the case, this is easier to grasp if you can see an example. So I suggest that you try the following.

1. In the file PrestigeCarsDimensionalWithDateDimension.pbix, click on the Sales table, and add two new measures containing the following DAX

```
Previous Month Sales =
CALCULATE(SUM([Sales[SalePrice]), PREVIOUSMONTH('Date'
[DateKey]))
```

```
Percent Of Previous Month Sales =
DIVIDE(SUM(Sales[SalePrice]), CALCULATE(SUM(Sales[
SalePrice]),PREVIOUSMONTH('Date'[DateKey])))
```

Figure 14-5 shows you the output that can be obtained by using the MonthFull field to give the month of a year (2021 in this example), the Total Sales measure to show the current month's sales alongside the Previous Month Sales and Percent Of Previous Month Sales measures based on the PREVIOUSMONTH() function.

MonthFull	Total Sales	Previous Month Sales	Percent Of Previous Month Sales
January	£569,450	£808,490	70.43%
February	£802,750	£569,450	140.97%
March	£1,058,270	£802,750	131.83%
April	£684,500	£1,058,270	64.68%
May	£1,828,390	£684,500	267.11%
June	£233,600	£1,828,390	12.78%
July	£861,400	£233,600	368.75%
August	£325,775	£861,400	37.82%
September	£470,250	£325,775	144.35%
October	£254,500	£470,250	54.12%
November	£455,700	£254,500	179.06%
December	£323,490	£455,700	70.99%

Figure 14-5. Comparing values with a previous time period

Once again the "magic" in the formula was the use of the CALCULATE() function. On this occasion a time intelligence function was used as a CALCULATE() modifier. Yet again this function required you to enter two elements:

- An aggregation to carry out (the sum of the Sale Price in this case).

- A filter to apply (the previous month in this example). Once again the function uses the date key from the date table to apply the time intelligence.

With these two parameters in place, DAX was able to take the time element used in the visualization (the month) and to say "give me the aggregation for the previous month."

As you can probably imagine, DAX does not limit you to comparing data month by month only. Indeed, it can help you to compare data for a wide range of time-spans. Once again (now that you have understood the basic principle), it is probably easiest to appreciate the related DAX functions as they are shown in Table 14-1.

Table 14-1. *DAX Date and Time Formulas to Return a Range of Dates*

Formula	Description	Example
PREVIOUSDAY()	Finds data for the previous day	CALCULATE(SUM(Sales[SalePrice]), PREVIOUSDAY('Date'[DateKey]))
PREVIOUSMONTH()	Finds data for the previous month	CALCULATE(SUM(Sales[SalePrice]), PREVIOUSMONTH('Date'[DateKey]))
PREVIOUSQUARTER()	Finds data for the previous quarter	CALCULATE(SUM(Sales[SalePrice]), PREVIOUSQUARTER('Date'[DateKey]))
PREVIOUSYEAR()	Finds data for the previous year	CALCULATE(SUM(Sales[SalePrice]), PREVIOUSYEAR('Date'[DateKey]))
NEXTDAY()	Finds the date for the following day	CALCULATE(SUM(Sales[SalePrice]), NEXTDAY('Date'[DateKey]))
NEXTMONTH()	Finds data for the following month	CALCULATE(SUM(Sales[SalePrice]), NEXTMONTH('Date'[DateKey]))
NEXTQUARTER()	Finds data for the following quarter	CALCULATE(SUM(Sales[SalePrice]), NEXTQUARTER('Date'[DateKey]))
NEXTYEAR()	Finds data for the following year	CALCULATE(SUM(Sales[SalePrice]), NEXTYEAR('Date'[DateKey]))

You need to be aware that the PREVIOUSMONTH() function only works as you want it to if the data that you are displaying is at month level. Figure 14-4 shows what happens if you mix up the time hierarchies. Figure 14-6 is displaying output at quarter level but using the PREVIOUSMONTH() function.

Quarter	Total Sales	Previous Month Sales	Percent Of Previous Month Sales
Q1	£2,430,470	£808,490	300.62%
Q2	£2,746,490	£1,058,270	259.53%
Q3	£1,657,425	£233,600	709.51%
Q4	£1,033,690	£470,250	219.82%

Figure 14-6. *Displaying data at the wrong level for a time intelligence function*

In this illustration the Previous Month Sales measure is only showing the sales for the first month of the previous quarter.

Similarly, PREVIOUSDAY() expects data to be displayed at day level, PREVIOUSQUARTER() expects data to be displayed at quarter level, and PREVIOUSYEAR() expects data to be displayed at year level.

Comparing Data over Any Time Period

Not all time-based data comparison uses only year on year metrics. To develop your knowledge of DAX time intelligence functions further, let's imagine that you want to compare current sales with sales for a previous period - be it a day, week, month, quarter, or year. Not only do you want to choose the *type of period* to compare current data with, but you also want to specify *how far back in time* you go when calculating the previous time period.

DAX can do all this in several different ways, but here is one fairly simple approach that returns the total vehicle sales for a previous time period. This is done using the DATEADD() function. The DATEADD() function lets you specify

- What time period you want to compare

- How far back (or forward) in time to "jump"

We'll start by calculating sales for the month before last

2. In the Sales table (still using the file PrestigeCarsDimensionalWithDateDimension.pbix) create a new measure.

3. Enter this DAX snippet

```
MonthBeforeLastSales =
CALCULATE(SUM(Sales[SalePrice]),
DATEADD('Date'[DateKey],-2,MONTH))
```

4. Press Enter or click the tick icon in the Formula Bar to finish the measure.

If you take a look at Figure 14-7, you will see how the sale price (at month level in this instance) will be compared to last month's sale price. In this example, the table shows only sales for 2021.

MonthAbbrAndYear	Total Sales	MonthBeforeLastSales
Jan 2021	£569,450	£111,890
Feb 2021	£802,750	£808,490
Mar 2021	£1,058,270	£569,450
Apr 2021	£684,500	£802,750
May 2021	£1,828,390	£1,058,270
Jun 2021	£233,600	£684,500
Jul 2021	£861,400	£1,828,390
Aug 2021	£325,775	£233,600
Sep 2021	£470,250	£861,400
Oct 2021	£254,500	£325,775
Nov 2021	£455,700	£470,250
Dec 2021	£323,490	£254,500
Total	**£7,868,075**	**£8,009,265**

***Figure 14-7.** Calculating metrics for a previous month*

So what the formula does is to aggregate the data in a column, but only for the month before the previous month, compared to the date field for each row.

The date elements that you can set as the third parameter of the DATEADD() function are

- Year

- Quarter

- Month

- Day

Although the function is called DATEADD(), it is nearly always (at least in my experience) used to look *back* in time rather than forward. So in this example you set the "timeshift" to a negative value, so that DAX would go back one month in time. You can, however, use positive numbers if you are looking at data from a past viewpoint and want to compare with data from later dates.

Once you have mastered this technique, you can extend and enhance a formula such as this to provide a multitude of metrics that will adapt to the time-based filters on tables and charts.

As a second example, try creating the following measures in the Sales table. They will calculate the total and average sale prices for the selected period(s) in the preceding quarter.

```
PreviousQuarterSales =
CALCULATE(SUM(Sales[SalePrice]),
DATEADD('Date'[DateKey], -1, QUARTER))
```

```
AverageSalePricePreviousQuarter =
CALCULATE(AVERAGE(Sales[SalePrice]),
DATEADD('Date'[DateKey],-1, QUARTER))
```

Alternatively perhaps you want to see the number of sales for the previous quarter relative to the date that is used to filter a visualization. This formula would be

```
NumberOfSalesPreviousQtr =
CALCULATE(
COUNT(Sales[ID]),
DATEADD('Date'[DateKey],-1,QUARTER)
)
```

Now that you have seen the principle, you are free to adapt it to your specific requirements. You can use any of the DAX aggregation functions that were described in the previous chapter, and can mix these with the four interval types (Year, Quarter, Month, and Day) that the DATEADD() function allows to deliver a truly wide-ranging set of time comparison metrics that will adapt automatically to the timespan of your Power BI Desktop visualization.

You need to be aware that the DATEADD() function only works if the data that you are displaying is set to the time element that you are using as the third parameter of the DATEADD() function. That is, if you are looking for data from a previous quarter, then you should be displaying data at quarter level. Figure 14-8 shows what happens if you mix up the time hierarchies (the data is filtered to display 2020 only).

MonthAbbrAndYear	SalePrice	PreviousMonthSales	PreviousQuarterSales
Jan 2020	£451,150.00	£69,500	£146,190
Feb 2020	£171,040.00	£451,150	£22,950
Mar 2020	£220,500.00	£171,040	£69,500
Apr 2020	£258,750.00	£220,500	£451,150
May 2020	£99,030.00	£258,750	£171,040
Jun 2020	£237,300.00	£99,030	£220,500
Jul 2020	£608,240.00	£237,300	£258,750
Aug 2020	£1,051,130.00	£608,240	£99,030
Sep 2020	£571,065.00	£1,051,130	£237,300
Oct 2020	£316,700.00	£571,065	£608,240
Nov 2020	£111,890.00	£316,700	£1,051,130
Dec 2020	£808,490.00	£111,890	£571,065
Total	£4,905,285.00	£4,166,295	£3,906,845

Quarter	SalePrice	PreviousQuarterSales	PreviousMonthSales
Q1	£842,690.00	£238,640	£691,690
Q2	£595,080.00	£842,690	£578,280
Q3	£2,230,435.00	£595,080	£1,896,670
Q4	£1,237,080.00	£2,230,435	£999,655
Total	£4,905,285.00	£3,906,845	£4,166,295

Figure 14-8. *Erroneous output when mixing up time hierachies*

I want to bring to your attention that DAX frequently provides alternate ways of calculating similar results. For instance,

```
1MonthBeforeSales =
CALCULATE(SUM(Sales[SalePrice]),
DATEADD('Date'[DateKey],-1,MONTH))
```

will give the same output as the measure Previous Month Sales that you created a few pages ago. So you are spoilt for choice when it comes to choosing the DAX that you want to use. In cases like this, when faced with an abundance of choice, there is often no "right" function - merely the one that you are happiest with that outputs a correct result.

Specifying Ranges of Dates

There may well be occasions when you want to set a specific data range in a time intelligence calculation. Suppose that you want to look at the total sales for the first fortnight in May and express sales for each day in May as a percentage of this specific date range. You could be doing this to see if all those public holidays combined with the prospect of summer vacation are stimulating vehicles sales.

Whatever the reason, you can use the DATESBETWEEN() function to set a specific date range in a CALCULATE() filter. Here is how.

1. In the Sales table (using the file PrestigeCarsDimensionalWithDateDimension.pbix), create a new measure like the following:

```
Holiday Sales =
SUM(Sales[SalePrice]) / CALCULATE(SUM(Sales[SalePrice]),
DATESBETWEEN('Date'[DateKey],
DATEVALUE("1/5/2022"),DATEVALUE("15/5/2022")))
```

2. Press Enter or click the tick mark icon to complete the measure definition.

3. In the Modeling ribbon, click the percentage icon to format this measure as a percentage.

You can see the kind of output that this produces in Figure 14-9.

MonthFull	DayOfMonth	Total Sales	Holiday Sales
May	1	£242,000	6.06%
May	9	£145,390	3.64%
May	10	£329,000	8.23%
May	12	£255,950	6.41%
May	13	£250,000	6.26%
May	16	£6,500	0.16%
May	19	£9,250	0.23%
May	20	£950	0.02%
May	21	£295,000	7.38%
May	22	£99,500	2.49%
May	23	£33,500	0.84%
May	24	£45,000	1.13%
May	26	£114,000	2.85%
May	27	£2,350	0.06%
Total		**£1,828,390**	**45.76%**

Figure 14-9. Setting a specific date range

The DATESBETWEEN() time intelligence function requires *three* parameters:

- Firstly the reference to the date key in the date table

- Secondly the start date in the range that you want to define

- Thirdly the end date in the range that you want to define

The entire function works like this:

- Calculate the total sales (for whatever the date display element that you have added to the visual).

- Divide this for the total sales for the specified date range.

Analyze Data as a Ratio over Time

Looking at how data aggregates over a specific stretch of time is only one of the ways that time intelligence can enable you to deliver time-based analysis. Sometimes you may well want to see how the sales for a chosen duration (day, week, month, or quarter) relate to the total sales for a period such as a year, quarter, or month. However, you don't want in this instance to "hard code" a date range, but define a formula that is more adaptable so that you can use it in different scenarios.

So in this example we will calculate the percentage of sales for each day that there were sales for Prestige Cars in the current year relative to the total sales for that year. This time, the DAX is flexible enough to use the year in any filters or in the data that is displayed.

The following formula adds greater flexibility through the use of the STARTOFYEAR() and ENDOFYEAR() functions that you saw briefly in the previous chapter. These take the date key column as their sole parameter and - as their names imply - deduce the beginning and end of the year.

Here is how it is done:

1. In the Sales table (using the file PrestigeCarsDimensionalWithDateDimension.pbix) create a new measure like the following:

```
Percent Of Year Sales =
SUM(Sales[SalePrice]) / CALCULATE(SUM(Sales[SalePrice]),
DATESBETWEEN('Date'[DateKey],
STARTOFYEAR('Date'[DateKey]),
ENDOFYEAR('Date'[DateKey]))
)
```

2. Press Enter or click the tick mark icon to complete the measure definition.

3. In the Modeling ribbon, click the percentage icon to format this measure as a percentage.

This formula was a little more complex than those that you have seen so far in this chapter. So let me explain it in a bit more detail. At its heart this formula consists of two main elements:

- The sales total for the time element that will be used in a visualization. This could be the day, week, month, or quarter, for instance.

- The total sales for the *entire year* that will be used in a visualization. This has to be calculated independently of the actual date element that will be displayed. Consequently the CALCULATE() function is used to extend the aggregation - the SUM() in this case - to the whole year. This is done by setting a range of dates as the filter for the aggregation. The date range is set using the DATESBETWEEN() function. This requires a lower threshold (defined by the STARTOFYEAR() function) and a higher threshold (defined by the ENDOFYEAR() function). This way the date range runs from the first to the last days of the year.

Finally the calculation divides the total for the specified time period by the total for the year to display the percentage that the period's sales represents.

So that you can see the outcome of your formula, you could create a matrix that contains the three following fields:

- FullYear

- FullMonth

- Total Sales

- Percent Of Year Sales

You should see something like Figure 14-10 if you have applied a page-level filter to restrict the FullYear to 2020 and 2021, for example.

FullYear	Total Sales	Percent Of Year Sales
⊟ **2020**		
January	£451,150	9.20%
February	£171,040	3.49%
March	£220,500	4.50%
April	£258,750	5.27%
May	£99,030	2.02%
June	£237,300	4.84%
July	£608,240	12.40%
August	£1,051,130	21.43%
September	£571,065	11.64%
October	£316,700	6.46%
November	£111,890	2.28%
December	£808,490	16.48%
⊟ **2021**		
January	£569,450	7.24%
February	£802,750	10.20%
March	£1,058,270	13.45%
April	£684,500	8.70%
May	£1,828,390	23.24%
June	£233,600	2.97%
July	£861,400	10.95%
August	£325,775	4.14%
September	£470,250	5.98%
October	£254,500	3.23%
November	£455,700	5.79%
December	£323,490	4.11%

Figure 14-10. *Displaying the sales per month as a percentage of the yearly sales*

You may be thinking that this was a lot of work just to get a percentage figure. Well, perhaps it is at first. Yet you can now capitalize on your effort and see how time intelligence is worth the effort. For instance, if you replace the MonthFull field in the table in Figure 14-9 with the Quarter field, you will instantly see the sales percentages for each quarter. Replace MonthFull with MonthAndDay and you will see the percentage of each day's sales relative to the entire year. This is because what you have done is to create an extremely supple and fluid formula that can adapt to any time segment. This is because what the formula does is to say "take the time segment (day, month, quarter, or year) and then find the date range for the whole year. Use this to calculate the total for the year-and then divide the figure for the time period by this total to display the percentage."

Figure 14-11 shows these two outputs.

FullYear	Total Sales	Percent Of Year Sales
⊟ **2020**		
Q1	£842,690	17.18%
Q2	£595,080	12.13%
Q3	£2,230,435	45.47%
Q4	£1,237,080	25.22%
⊟ **2021**		
Q1	£2,430,470	30.89%
Q2	£2,746,490	34.91%
Q3	£1,657,425	21.07%
Q4	£1,033,690	13.14%

FullYear	Total Sales	Percent Of Year Sales
⊟ **2020**		
Jan 1	£349,500	7.12%
Jan 7	£32,050	0.65%
Jan 9	£12,650	0.26%
Jan 22	£56,950	1.16%
Feb 16	£39,500	0.81%
Feb 17	£71,890	1.47%
Feb 22	£56,000	1.14%
Feb 28	£3,650	0.07%
Mar 20	£220,500	4.50%
Apr 5	£102,950	2.10%
Apr 30	£155,800	3.18%
May 30	£99,030	2.02%
Jun 15	£237,300	4.84%
Jul 1	£15,650	0.32%
Jul 6	£12,500	0.25%
Jul 25	£486,000	9.91%
Jul 26	£94,090	1.92%
Aug 2	£167,350	3.41%
Aug 9	£99,500	2.03%
Aug 10	£12,500	0.25%
Aug 12	£61,500	1.25%
Aug 13	£196,100	4.00%
Aug 19	£64,500	1.31%
Aug 21	£76,500	1.56%
Aug 22	£45,900	0.94%
Aug 23	£125,000	2.55%

Figure 14-11. *Displaying the sales per day and per quarter as a percentage of the yearly sales*

When setting comparative date ranges, you may also need to apply the following functions:

- STARTOFQUARTER()

- ENDOFQUARTER()

- STARTOFMONTH()

- ENDOFMONTH()

These functions allow you to define data ranges that take the current date and extrapolate backwards and forwards to the start and end of quarters and years.

These functions are virtually identical to the STARTOFYEAR() and ENDOFYEAR() functions. So merely copying the Percent Of Year Sales measure would scarcely add any value to your learning curve. Instead, then I want to extend these functions a little further by using them to display the month or quarter totals *only if there were sales on a date*. This avoids having a lot of blank rows in the output for days when no cars were sold.

The formula that does this is (at the month level)

```
Sales for Current Month =
IF(NOT ISBLANK(SUM(Sales[SalePrice])),
CALCULATE(SUM(Sales[SalePrice]),
DATESBETWEEN('Date'[DateKey],
STARTOFMONTH('Date'[DateKey]),
ENDOFMONTH('Date'[DateKey])))
)
```

At the quarter level the formula is

```
Sales for Current Quarter =
IF(NOT ISBLANK(SUM(Sales[SalePrice])),
CALCULATE(SUM(Sales[SalePrice]),
DATESBETWEEN('Date'[DateKey],
STARTOFQUARTER('Date'[DateKey]),
ENDOFQUARTER('Date'[DateKey])))
)
```

You can see these functions in action in Figure 14-12.

FullYear	Total Sales	Sales for Current Month	Sales for Current Quarter
⊟ **2020**			
Jan 1	£349,500	$451,150	842690
Jan 7	£32,050	$451,150	842690
Jan 9	£12,650	$451,150	842690
Jan 22	£56,950	$451,150	842690
Feb 16	£39,500	$171,040	842690
Feb 17	£71,890	$171,040	842690
Feb 22	£56,000	$171,040	842690
Feb 28	£3,650	$171,040	842690
Mar 20	£220,500	$220,500	842690
Apr 5	£102,950	$258,750	595080
Apr 30	£155,800	$258,750	595080
May 30	£99,030	$99,030	595080
Jun 15	£237,300	$237,300	595080
Jul 1	£15,650	$608,240	2230435
Jul 6	£12,500	$608,240	2230435
Jul 25	£486,000	$608,240	2230435
Jul 26	£94,090	$608,240	2230435
Aug 2	£167,350	$1,051,130	2230435
Aug 9	£99,500	$1,051,130	2230435
Aug 10	£12,500	$1,051,130	2230435
Aug 12	£61,500	$1,051,130	2230435
Aug 13	£196,100	$1,051,130	2230435
Aug 19	£64,500	$1,051,130	2230435
Aug 21	£76,500	$1,051,130	2230435
Aug 22	£45,900	$1,051,130	2230435
Aug 23	£125,000	$1,051,130	2230435

Figure 14-12. Using the STARTOFMONTH(), ENDOFMONTH(), STARTOFQUARTER(), and ENDOFQUARTER() functions

In these metrics, the total for the month is calculated by

```
CALCULATE(SUM([SalePrice]),
DATESBETWEEN('Date'[DateKey],STARTOFMONTH('Date'[DateKey])
,ENDOFMONTH('Date'[DateKey])))
```

However, to prevent empty rows cluttering up the output, the CALCULATE() function is wrapped inside the following DAX:

```
NOT ISBLANK(SUM([SalePrice])), ..... )
```

This short extension to the code tests whether the total sales are empty using the ISBLANK() function. The logic is reversed using the NOT operator so that only if there are sales for a day the month (and quarter) calculations are applied.

Extending Core Time Intelligence Functions

The measures that you created in the previous section showed you that time intelligence functions occasionally need to be extended slightly to ensure that the output is either tailored to display only relevant data or easier to read or both.

As real-world analytics often requires measures that, while being based on core functions, need to be wrapped in display logic and error handling, I am going to make the final few time intelligence examples in this chapter a little more verbose (notice that I did not say complicated). This way you can see, as well, some of the techniques that you may have to apply when writing your own time analytics in DAX.

Comparing Data from Previous Years

The following two measures (YearOnYearDelta and YearOnYearDeltaPercent) calculate the increase or decrease in sales compared to a previous year and also that change expressed as a percentage. These measures extend the logic of the last few formulas using functions that you have already met. I will presume that now, after a large chapter on DAX time intelligence that you do not really need a step by step explanation of how to enter a formula. So I will only present and then explain the code from now on. The code to add YearOnYearDelta as a new measure to the Sales table is as follows:

```
YearOnYearDelta=
IF(

ISBLANK(
SUM(Sales[SalePrice])
),
BLANK(),
```

```
IF(
ISBLANK(
CALCULATE(
SUM(Sales[SalePrice]),
DATEADD('Date'[DateKey], -1, YEAR)
)
),
BLANK(),
SUM(Sales[SalePrice])
- CALCULATE(
SUM(Sales[SalePrice]),
DATEADD('Date'[DateKey], -1, YEAR)
)
)
)
```

The code to add YearOnYearDeltaPercent as a new measure to the Sales table is

```
YearOnYearDeltaPercent=
IF(
ISBLANK(
SUM(Sales[SalePrice])
),
BLANK(),
IF(
ISBLANK(
CALCULATE(
SUM(Sales[SalePrice]),
DATEADD('Date'[DateKey], -1, YEAR)
)
),
BLANK(),
(
SUM(Sales[SalePrice])
- CALCULATE(
SUM(Sales[SalePrice]),
```

```
DATEADD('Date'[DateKey], -1, YEAR)
)
)
/CALCULATE(
SUM(Sales[SalePrice]),
DATEADD('Date'[DateKey], -1, YEAR)
)
)
)
```

These two formulas are a lot easier than they look, believe me. The formula for YearOnYearDelta is really only

```
SUM(Stock[SalePrice])
- CALCULATE(SUM(Sales[SalePrice]), DATEADD(Date[DateKey], -1, YEAR))
```

All the code says is "Subtract last year's sales from this year's sales." Everything else is wrapper code to prevent a calculation if either this year's or last year's data is zero.

Equally, for the formula *YearOnYearDeltaPercent*, the core code is this:

```
SUM(Sales[SalePrice]) - CALCULATE(SUM(Sales[SalePrice]),DATEADD('Date'[Date
Key], -1, YEAR)) / CALCULATE(SUM(Sales[SalePrice]),DATEADD('Date'[DateKey],
-1, YEAR))
```

In other words, "Subtract last year's sales from this year's sales and divide by last year's sales." Everything else in the complete formula that is initially given in full above exists to prevent divide-by-zero errors or unwanted results for the first year where there is no previous year's data.

The logic "wrapper" around the core formula uses two functions that you saw in Chapter 7, but it is worth taking another look at them here. They are

- ISBLANK() — This function tests if a calculation returns nothing and allows you to specify what to do if this happens. This is a bit like an IF function that only tests for blank data.

- BLANK() — Returns a blank (or Null). This is useful for overriding unwanted results and replacing them with a blank.

Using these functions lets you handle the case where there is no data for a previous time period. As you are using the ISBLANK() function to test for inexistent data, you are able to replace any missing data with a BLANK() - rather than letting Power BI Desktop display an unsightly error.

We have seen a couple of fairly complex formulas in a short section. So I think that it is a good idea to see how they look when you apply them. In this case, I will use a Power BI Desktop table to show the results, as you can see in Figure 14-13. As you can see the appropriate formats have been applied to each metric to enhance readability. This data is for 2022.

MonthFull	Total Sales	PreviousYearSales	YearOnYearDelta	YearOnYearDeltaPercent
January	£817,710	£569,450	£248,260	43.60%
February	£1,166,900	£802,750	£364,150	45.36%
March	£756,390	£1,058,270	-£301,880	-28.53%
April	£1,070,135	£684,500	£385,635	56.34%
May	£755,125	£1,828,390	-£1,073,265	-58.70%
June	£194,525	£233,600	-£39,075	-16.73%
July	£932,980	£861,400	£71,580	8.31%
August	£631,040	£325,775	£305,265	93.70%
September	£504,940	£470,250	£34,690	7.38%
October	£405,390	£254,500	£150,890	59.29%
November	£28,990	£455,700	-£426,710	-93.64%
December	£519,490	£323,490	£196,000	60.59%
Total	**£7,783,615**	**£7,868,075**	**-£84,460**	**-1.07%**

Figure 14-13. Power BI Desktop output for year on year comparisons

Once again these formulas only scratch the surface of the myriad possibilities that DAX has on offer. However you can adapt them to create comparisons by quarter, month, or day if you prefer simply by changing the specified time interval from YEAR to QUARTER, MONTH, or DAY.

Rolling Aggregations over a Period of Time

We are now getting into the arena of more complex DAX formulas. So, since returning the rolling sum (or average) of a specified period to date necessitates several DAX functions and some in-depth nesting of these functions, I will take this as an example of the kind of more complicated DAX measure that you are likely to need in real life. I will begin by outlining some of the functions that are used to deliver a result that is reliable and efficient:

- DATESBETWEEN() — Lets you select a range of dates. The three parameters are the date key field from the date table, then the starting date, and then the ending date.

- FIRSTDATE() — Allows you to get the first date from a range. Since we are using this momentarily to go back a defined number of months, it will get the first day of the month.

- LASTDATE() — Allows you to get the last date from a range. Since we are using this momentarily to go back a defined number of months, it will get the last day of the month.

You can now create two measures in the Sales table (3MonthsToDate and Previous3Months) using some additional logic to ensure that only blank cells are returned if there is no previous year's data using the following formulas:

```
3MonthsToDate =
IF(
ISBLANK(SUM(Sales[SalePrice])),
BLANK(),
CALCULATE(
SUM(Sales [SalePrice]),
DATESINPERIOD(
'Date'[DateKey],
LASTDATE('Date'[DateKey]),-3,MONTH)
)
)
```

```
Previous3Months =
IF(
ISBLANK(
CALCULATE(
SUM(Sales [SalePrice]),
DATEADD('Date'[DateKey],-1,MONTH)
)
),
BLANK(),
CALCULATE(
SUM(Sales [SalePrice]),
DATESBETWEEN(
'Date'[DateKey],
FIRSTDATE(DATEADD('Date'[DateKey],
-3,MONTH)),
LASTDATE(DATEADD('Date'[DateKey],
-1,MONTH))
)
)
)
```

The formula 3MonthsToDate essentially evaluates the code that is in boldface. This
says, "Add up the sales for a time period ranging from three months ago to now." The IF
function detects if there are sales for the current date, and if there are none (ISBLANK),
then the calculation is not attempted, and a blank (or null or empty) value is returned.

The formula *Previous3Months* is pretty similar, except that the time period uses the
DATESBETWEEN() function to set a range of dates — from the first day of the month
three months ago to the last date in the preceding month.

MonthAbbrAndYearAbbr	Total Sales	3MonthsToDate	Previous3Months
Jan 21	£569,450	£1,489,830	£1,237,080
Feb 21	£802,750	£2,180,690	£1,489,830
Mar 21	£1,058,270	£2,430,470	£2,180,690
Apr 21	£684,500	£2,545,520	£2,430,470
May 21	£1,828,390	£3,571,160	£2,545,520
Jun 21	£233,600	£2,746,490	£3,571,160
Jul 21	£861,400	£2,923,390	£2,746,490
Aug 21	£325,775	£1,420,775	£2,923,390
Sep 21	£470,250	£1,657,425	£1,420,775
Oct 21	£254,500	£1,050,525	£1,657,425
Nov 21	£455,700	£1,180,450	£1,050,525
Dec 21	£323,490	£1,033,690	£1,180,450
Jan 22	£817,710	£1,596,900	£1,033,690
Feb 22	£1,166,900	£2,308,100	£1,596,900
Mar 22	£756,390	£2,741,000	£2,308,100
Apr 22	£1,070,135	£2,993,425	£2,741,000
May 22	£755,125	£2,581,650	£2,993,425
Jun 22	£194,525	£2,019,785	£2,581,650
Jul 22	£932,980	£1,882,630	£2,019,785
Aug 22	£631,040	£1,758,545	£1,882,630
Sep 22	£504,940	£2,068,960	£1,758,545
Oct 22	£405,390	£1,541,370	£2,068,960
Nov 22	£28,990	£939,320	£1,541,370
Dec 22	£519,490	£953,870	£939,320
Total	**£15,651,690**	**£953,870**	**£16,369,280**

Figure 14-14. *Rolling time period calculations*

Which DAX Functions to Use for Comparison over Time?

In the course of this chapter, you have seen several DAX functions that you can also use when comparing data over time. A quick recapitulation of the core timespan definition functions is given in Table 14-2.

Table 14-2. *DAX Date and Time Formulas to Compare Values over Time*

Formula	Description
PARALLELPERIOD()	Finds dates from a "parallel" timeframe defined by a certain number of set intervals. The first parameter is the source data, the second is the number of years, quarters, months, or days, and the third is the definition of the interval – years, quarters, months, or days. A positive number of intervals looks forward in time, and a negative number goes backward in time
SAMEPERIODLASTYEAR()	Finds the date(s) for the same time range one year before
DATEADD()	Used to return data from a past or future period in time compared to a specified date. The date difference can be in years, quarters, months, or days
DATESBETWEEN()	Calculates a list of dates between two dates. The first parameter is the start date, and the second parameter is the end date
DATESINPERIOD()	Calculates a list of dates beginning with a start date for a specified period

Defining Relative Time Periods in DAX

DAX contains several functions that you can use to return relative dates when aggregating data over time. These are recapitulated in Table 14-3.

Table 14-3. *DAX Date and Time Formulas to Return a Date*

Formula	Description	Example
FIRSTDATE()	Finds the first date that an event took place – such as the first day in a month that there was a sale	FIRSTDATE(Sales[Invoice Date])
LASTDATE()	Finds the last date that an event took place – such as the last day in a month that there was a sale	LASTDATE(Sales[Invoice Date])
FIRSTNONBLANK()	Returns the first value that is not blank (empty) in a range of data. The first parameter is the column to return; the second is the column to iterate down until a blank cell is found	FIRSTNONBLANK(Sales[SalePrice], Date(DateKey))
LASTNONBLANK()	The first parameter is the column to return, the second is the column to iterate down until the last non-blank cell is found	LASTNONBLANK(Sales[Sale Price], Date(DateKey))

Time Intelligence Without a Date Dimension

It is technically possible to apply some time intelligence functions to data and datetime columns in tables where the data model does not have a date dimension. However I find this can rapidly become complex and confusing. So I will not be adding examples of this approach in this chapter. My core advice is - if you want to analyze over time, add a date dimension and use DAX time intelligence the way that it was designed to be used. You will make your DAX programming life much easier.

Conclusion

This chapter has taken you on a short tour of some of the ways that you can use DAX in Power BI Desktop to extract meaning from your data by analyzing its evolution over time. This has, in effect, been an introduction to DAX time intelligence functions.

You have examined a set of "To Date" functions that can isolate aggregate metrics to the end of the current month, quarter, or year. You also saw how to extract data from previous time periods in several ways, depending on the specific requirement.

Then you learned how to use ranges of dates when delivering analytics over time. These date ranges can be relative to a current date or fixed. Finally you saw how to extend time intelligence functions to ensure that they only displayed the required data and handled time period limits.

All the measures that you saw in this chapter can be found in the Power BI Desktop file PrestigeCarsDimensionalWithTimeIntelligence.pbix. This file is included in the downloadable sample data for this book.

I am afraid that explaining all the possibilities of DAX time intelligence would take an entire book, so all I wanted to do in this chapter and the previous one was explain how you can use DAX time intelligence in a handful of useful calculations. I sincerely hope that this brief overview will help you on your road to mastery of DAX time intelligence and that you will be able to apply these formulas to deliver some powerful and cogent analytics over time using Power BI Desktop.

CHAPTER 15

DAX Variables

As you have progressed through the chapters of this book, you have seen how DAX code can increase in complexity as you have to solve more interesting analytical challenges. While there is nothing inherently wrong with intricate code structures that deliver a valid result, long and convoluted DAX measures can be both hard to write and even harder to debug and maintain.

So the developers of DAX have come up with an elegant solution to handling more advanced coding challenges. This is to use DAX variables. Variables in DAX

- Simplify DAX development

- Clarify complex measures

- Make DAX easier to read and understand

- Avoid repeating code and duplicating calculations through making parts of the code reusable

- Make DAX run faster in many circumstances

- Can assist in measure definition and writing DAX through helping you to break down a coding challenge into component parts

In this chapter you will learn how to add variables to DAX. Firstly you will learn to enhance the kinds of measures that you have written in previous chapters, and then you will move on to applying variables to enable DAX to produce results that are much easier to deliver using variables.

To continue building on the code - and experience - that you have gained from the previous chapters, I will be using the Power BI Desktop file PrestigeCarsDimensionalWithTimeIntelligence.pbix as the starting point for learning about DAX variables in this chapter. You can download this file from the Apress website as part of the support material for this book.

© Adam Aspin 2023
A. Aspin, *Pro DAX and Data Modeling in Power BI*, https://doi.org/10.1007/978-1-4842-8995-2_15

All About Variables

Before seeing how variables can be used in DAX, you need to know few basic ground rules as far as variables are concerned.

Variable Usage

Variables are used in DAX to store parts of a measure. They can be used anywhere in your DAX code where you would otherwise employ the element that you set the variable to contain. This means that variables can be substituted for and can contain

- Strings

- Numbers

- Formulas

- Tables (which you will learn about in the following chapter)

Variable Names

DAX variables must be introduced in a specific way. This is called *declaring* variables. Essentially, this means giving a variable a name. There are a few simple rules to respect when declaring variables. Variable names

- Must begin with a letter or an underscore

- May contain numbers after the first letter

- Can be any length (but you will probably want to keep them short yet meaningful)

- Cannot contain spaces

- Must be made up of alphanumeric characters only – that is, they cannot contain special characters or punctuation

- Cannot be wrapped in quotes (single or double) or parentheses, square brackets, or braces

- Can be in any mixture of uppercase and lowercase characters and underscores

Fortunately you do not have to memorize these rules at any cost, as Power BI Desktop will indicate in the formula bar if a variable name is invalid. It does this by adding a red wavy line under the variable name.

You can name your variables any way you want. Some DAX developers prefer to start all variable names with an underscore (or even a double underscore). Others apply no rules (or naming convention as this is also known) at all. I will be using the following convention in this book:

- n_ begins a variable that contains a *number* to be used in the formula (n is for *numeric constant*).

- s_ begins a variable that contains a *text* to be used in the formula (s is for *string*).

- c_ begins a variable that contains a DAX *formula* snippet to be used in the formula (c is for *calculation*).

- t_ begins a variable that contains a *table* to be used in the formula (t is for *table*).

I prefer to be rigorous and use a naming convention as I find that it makes the DAX clearer. When a variable is instantly recognizable, you know that you are not looking at a table or field but a variable. You are, of course, free to use any naming convention or none at all in your DAX.

Creating Variables

As I mentioned above, variables have to be *declared*. This is another way of saying that DAX needs to know that you want to use a variable, and consequently you have to introduce it into the DAX code in a specific way.

This is, fortunately, blissfully simple. All you have to do is add the keyword VAR (in upper- or lowercase) before a variable name. You will see this very shortly.

All you need to remember is that a variable must contain the following four elements:

- The VAR keyword (in uppercase or lowercase or a mixture of both – DAX is very forgiving)

- A valid variable name

- The equals sign

- The contents of the variable

You cannot partially declare a variable by not assigning a value to it. You must add all the elements or nothing.

Using Variables

A really important point to remember is that once you start using variables, you will have to add the RETURN keyword to your DAX to conclude the use of a variable (or even multiple variables).

So, at a minimum, variables in DAX

- Begin with VAR

- End with RETURN

One important point (that is worth noting by Power BI users who have programming experience) is that variables are, really, *constants*. That is once their value is set, it *cannot be altered* in the course of running the DAX measure that contains the variable.

Moreover, variables are limited to the measure in which they are written. So you cannot declare a variable inside one measure and use it in another.

You do *not*, however, have to use a variable once it has been declared.

Variable Output

Once a variable has been processed, it will hand back a value to the measure that contains it. This value can be

- A scalar value (one single value which can be a text, date, Boolean value or a number). This value can be calculated or simply set.

- A table.

The art and science of using variables is to ensure that the return value from a variable (the output) matches what you need in the rest of the measure.

Basic Variable Use

That is enough theory. It is now time to give you a simple idea of how variables are implemented. As variables are new and different, I will explain the first one step by step.

Suppose you want to create a measure that gives a projected ten percent increase in sales. Here is one way of writing this using a variable in a measure added to the Sales table:

1. Open the file PrestigeCarsDimensionalWithTimeIntelligence.pbix, and select the Sales table in the Fields Pane.

2. Add a new measure.

3. Enter **Sales Projection** to the left of the equals sign.

4. Press Shift-Enter to begin a new line.

5. Enter the VAR keyword in the Formula bar, followed by the variable name (**n_PercentageIncrease** in this example).

6. Add an equals sign then the value that the variable will contain. This is **0.1** in this example.

7. Press Shift-Enter to begin a new line.

8. Add the **RETURN** keyword.

9. Press Shift-Enter to begin a new line.

10. Add the following DAX:

    ```
    ([Total Sales] * n_PercentageIncrease
    ```

11. Confirm the formula. The DAX that you have created should look like this:

    ```
    Sales Projection =
    VAR n_PercentageIncrease = 0.1

    RETURN
    CALCULATE([Total Sales] * n_PercentageIncrease)
    ```

    Of course, this measure could have been written as

    ```
    Sales Projection = CALCULATE([Total Sales] * 0.1)
    ```

However the point here is to provide a gentle introduction to variables. The way that the variable is used is explained graphically in Figure 15-1.

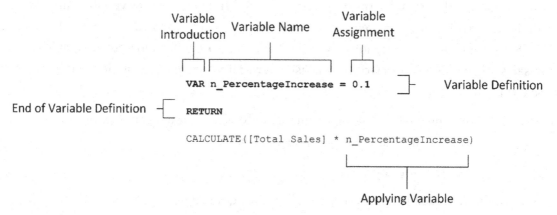

Figure 15-1. *Variable implementation*

In this specific case the advantage of using a variable is that the percentage increase is clearly visible at the start of the formula and not buried deep inside a potentially complex piece of code. This is one way that using variables in DAX can add clarity.

Note   Variable names are *not* case-sensitive. So you can declare a variable in uppercase and use it later in lowercase, and it will still work. However this will defeat one of the advantages of using variables which is to add clarity to the DAX.

DAX containing variables can be written on one line if you want. However, as one of the reasons to use variables is to make DAX easier to read, it seems a shame to write DAX like this:

```
Sales Projection = var n_PercentageIncrease = 0.1 RETURN CALCULATE([Total
Sales] * n_PercentageIncrease)
```

Variables and Intellisense

Once you have declared a variable in DAX, you never need type its name again. This is because the variable will now appear in the Intellisense popup in the Formula bar.

So, if you want to refer in a measure to a variable that you have previously declared, all you have to do is to start typing a few letters contained in the variable name and it will appear, ready to be selected in the popup list. You can see this in Figure 15-2.

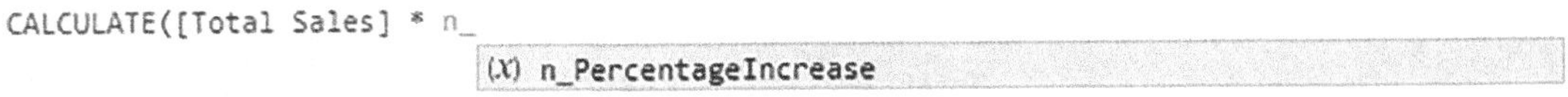

```
CALCULATE([Total Sales] * n_
```

(X) n_PercentageIncrease

Figure 15-2. *Intellisense prompts for variables*

As you can see from Figure 15-2, all DAX variables are preceded by a small (X) symbol. This symbol is the same whatever the variable contains.

Implementing variables from the popup can be another reason for employing a valid naming convention when using variables in DAX. This also underlines the fact that giving variables meaningful names will help you in your DAX coding.

Basic Variable Assignment

Assignment is the art of saying what a variable will hold. As I mentioned earlier this can be a number, a text (or string if you prefer), a calculation, or a table. In the interests of completeness, let's look briefly at assigning texts, calculations, and tables as you saw how to assign a number in the previous example.

Assigning Texts to Variables

Declaring a variable that will contain text is outrageously simple. You can see this in the following DAX snippet where a text variable is declared and then used as the text element of a CALCULATE() filter. This measure could be added to the Geography table.

```
US Sales To Total =
VAR s_US = "United States"

RETURN
DIVIDE(CALCULATE(SUM(Sales[SalePrice]), 'Geography'[CountryName] = s_US),
[Total Sales])
```

You can see the output for this measure in Figure 15-3.

FullYear	US Sales To Total
2019	11.65%
2020	6.00%
2021	4.73%
2022	14.95%
Total	**9.08%**

Figure 15-3. *Using variables with text values*

Text-based variables can be extended to deliver more subtle filtering (and use of text). So, for instance, you could declare a text variable that contains a list of elements that you want to use in a filter. The following piece of DAX shows this in action.

```
Core Europe Sales To Total =
VAR t_CoreEurope = {"Spain", "France", "Belgium"}

RETURN
DIVIDE(CALCULATE(SUM(Sales[SalePrice]), 'Geography'[CountryName] IN s_
CoreEurope), [Total Sales])
```

In cases like this the variable includes the braces that are needed for a list used inside an IN filter inside CALCULATE().

Text variables have a myriad of possible applications. Here you have seen only a couple of simple examples.

Assigning Calculations to Variables

One of the more frequent uses of variables in DAX is to use them to hold the output from a calculation. The calculation declared as a variable can be as simple or as complex as you want it to be.

As the variable "represents" a calculation, it has the same limitations as any other calculation. So, if you are writing a DAX *measure* using a variable, the calculation must at the very least contain an aggregate function as variables - like measures - cannot refer to a "cell" in a row if the variable is used inside a measure. Equally, variables containing calculations inside calculated columns must respect the limitations of calculated columns.

This is probably easy to understand if you see an example. So take a look at the following DAX code that creates a measure in the Sales table.

```
UK To All Sales % =
VAR c_UKSales =   CALCULATE(SUM(Sales[SalePrice]),
                  'Geography'[CountryName] = "United Kingdom")

RETURN
DIVIDE(c_UKSales, [Total Sales])
```

You can see the output from this measure in Figure 15-4.

FullYear	UK To All Sales %
2019	77.20%
2020	52.47%
2021	35.07%
2022	35.73%
Total	**41.91%**

Figure 15-4. *Using variables for calculations*

In this example the numerator used by the DIVIDE() function is the output from a CALCULATE() function. This output is stored in the variable c_UKSales which is passed as the first parameter of DIVIDE(). This is shown graphically in Figure 15-5.

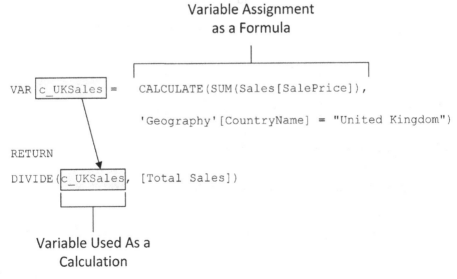

Figure 15-5. *Variables for calculations explained*

Isolating parts of a more complex DAX function like this greatly increases the clarity of the code - even when the measure is as simple as this one is. When it comes to more complex measures using variables as placeholders for calculation elements can really make the DAX easier to understand.

Assigning Tables to Variables

DAX variables can also be assigned tables as their content. While you will be learning more about tables in the next chapter, one simple - and frequently used - application of table variables is to separate out FILTER() functions. As you saw in Chapter 11, the FILTER() function returns a table.

What the following code does is to define a FILTER() function that will select one make of car. This is then used to filter the output from a CALCULATE() function - which itself is used in a DIVIDE() function. Here is the measure (also in the Sales table).

```
Aston Martin To Total Sales % =
VAR t_AstonMartin =
        FILTER('Vehicle', 'Vehicle'[Make] = "Aston Martin")

RETURN
DIVIDE(CALCULATE([Total Sales], t_AstonMartin), [Total Sales])
```

You can see the output from this measure in Figure 15-6.

FullYear	Aston Martin To Total Sales %
2019	31.01%
2020	24.29%
2021	18.23%
2022	26.16%
Total	**23.22%**

Figure 15-6. *Assigning a table to a variable*

Figure 15-7 explains this approach more graphically.

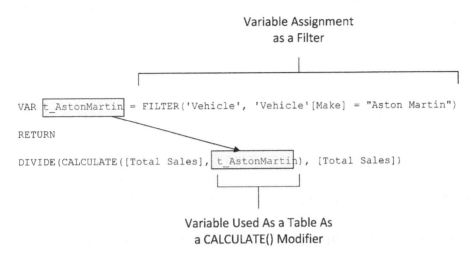

Figure 15-7. *Assigning a table to a variable explained*

This piece of DAX also introduced you to the fact that a table can be used as a CALCULATE() filter.

What to RETURN

Up until now in this chapter we have used variables where the output - as defined by what follows the RETURN keyword - has nearly always been a calculation. This is not, however, the only way of delivering output from a variable in a measure.

Another solution is to declare a variable that will contain the value that is returned by a calculation, and use this in the RETURN statement. As ever, this is easiest to understand through an example. So take a look at the following DAX snippet. Here, a variable is declared (n_Total) that adds together the previous two variables. It is this variable that is then used in the RETURN statement to deliver the output from the following formula that has been added as a measure to the Sales table.

```
Spares and Labor Cost Analysis =
VAR c_spares = SUM(Sales[SpareParts])
VAR c_Labor = SUM(Sales[LaborCost])
VAR n_Total = c_Labor + c_spares

RETURN
c_Total
```

You can see the output from this measure in Figure 15-8.

MonthAbbrAndYear	Total Sales	Spares and Labor Cost Analysis
Jan 2019	£304,950	£13,235
Feb 2019	£31,000	£1,910
Mar 2019	£155,000	£9,250
Apr 2019	£145,100	£5,565
May 2019	£373,450	£16,450
Jun 2019	£22,950	£1,050

Figure 15-8. *Using a variable as the RETURN element*

In this particular DAX snippet the third variable, n_Total, only exists to store the output from the calculation that uses the two previous variables. This output is returned as the output from the measure when it is used in a visual.

It is entirely up to you as to which approach you take to define the output from a DAX variable.

Multiple Variables in a Measure

Now that you are familiar with the basic approach to variables, we can develop their application a little further.

One of the first things you need to be aware of is that there is no limit to the number of variables that you can use inside a measure. So there is nothing to stop you defining a series of variables - as you can see in the following example (this time added to the Geography table).

```
Multi Country Filter =
VAR s_Country1 = "Germany"
VAR s_Country2 = "Switzerland"

RETURN
CALCULATE([TotalSales]),
'Geography'[CountryName] = s_Country1 || 'Geography'[CountryName] = s_
Country2)
```

This measure gives the output that is displayed in Figure 15-9.

FullYear	Multi Country Filter
2019	£11,500
2020	£613,340
2021	£680,300
2022	£407,315
Total	**£1,712,455**

Figure 15-9. *Using multiple variables in a measure*

The key points to remember at this juncture are

- You can define as many variables as you like

- You do *not* have to use variables once declared

- You can use variables in any order once they have been declared

One core takeaway is that you *must* declare a variable (and assign a value to it) *before* you can use it or Intellisense will indicate an error - and the DAX will not work.

To make this clearer, take a look at the following piece of DAX. This measure *will not work* as the variables n_Labor, n_spares, and n_Delivery are declared *after* being used.

```
Indirect Cost Analysis =
VAR c_Output = n_Labor + n_spares + n_Delivery
VAR n_spares = SUM(Sales[SpareParts])
VAR n_Labor = SUM(Sales[LaborCost])
VAR n_Delivery = SUM(Sales[DeliveryCharge])
RETURN
c_Output
```

This piece of DAX will, by contrast, work perfectly as no variable is used before it has been declared. This measure has been added to the Sales table.

```
Indirect Cost Analysis =
VAR n_spares = SUM(Sales[SpareParts])
VAR n_Labor = SUM(Sales[LaborCost])
VAR n_Delivery = SUM(Sales[DeliveryCharge])
VAR c_Output = n_Labor + n_spares + n_Delivery

RETURN
c_Output
```

You can see the output from the measure in Figure 15-10.

FullYear	Indirect Cost Analysis
2019	£81,330
2020	£259,405
2021	£355,865
2022	£392,372
Total	**£1,088,972**

Figure 15-10. *Declaring variables before use*

It is worth noting that Intellisense will help you here, as it will only display variables that it can use at any specific part of the code. That is, only variables that have already been declared will appear in the Intellisense popup in the Formula bar.

Variable Reuse Inside a Measure

Simplifying DAX is one excellent reason to use variables. Another equally good motivation to replace calculations (or text or numbers) with variables is to enable code reuse.

As an example, imagine that you have to create a measure that calculates the percentage difference between this month's and last month's sales. Assuming that you do not already have a measure that calculates last month's sales, the DAX that you will write will need to use the calculation for last month's sales twice:

- Once to deduct last month's sales from this month's sales to obtain the difference

- Once to divide the difference you just calculated by last month's sales

This can be done by repeating the piece of code that calculates last month's sales, but it makes for long-winded code that is messier than it needs to be. Instead you can declare a variable to calculate last month's sales and then use the variable twice instead of repeating the formula. This is the DAX to do this.

```
Month On Month % =
VAR c_PreviousMonthSales =
    CALCULATE([Total Sales], PREVIOUSMONTH('Date'[DateKey]))
RETURN

DIVIDE(([Total Sales] - c_PreviousMonthSales), c_PreviousMonthSales)
```

You can see the output from the measure - added to the Sales table - in Figure 15-11. I have included the figure for the previous month's sales as a sense-check.

MonthAbbrAndYear	Total Sales	Previous Month Sales	Month On Month %
Jan 2019	£304,950		
Feb 2019	£31,000	£304,950	-89.83%
Mar 2019	£155,000	£31,000	400.00%
Apr 2019	£145,100	£155,000	-6.39%
May 2019	£373,450	£145,100	157.37%
Jun 2019	£22,950	£373,450	-93.85%
Jul 2019	£107,185	£22,950	367.04%
Aug 2019	£5,500	£107,185	-94.87%
Sep 2019	£37,200	£5,500	576.36%
Oct 2019	£146,190	£37,200	292.98%
Nov 2019	£22,950	£146,190	-84.30%
Dec 2019	£69,500	£22,950	202.83%
Jan 2020	£451,150	£69,500	549.14%
Feb 2020	£171,040	£451,150	-62.09%
Mar 2020	£220,500	£171,040	28.92%
Apr 2020	£258,750	£220,500	17.35%
May 2020	£99,030	£258,750	-61.73%

Figure 15-11. *Output when reusing variables*

To show how variables are used, a more graphical representation of the code is given in Figure 15-12.

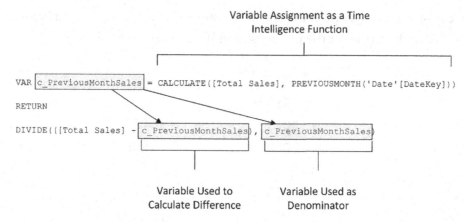

Figure 15-12. *Reusing variables*

To encourage you to use variables, it is worth remembering that

- You can reuse variables any number of times inside a measure

- Reusing variables can make measures run faster

Variables in Calculated Columns

I promised you earlier in this chapter that we would take a look at using variables in calculated columns. After all, DAX is used in calculated columns as well as in measures. Consequently variables can be applied to both.

When using variables the key thing to remember is the type of calculation that you are creating, as variables do not alter the fundamentals of DAX and how it is applied. So if you are using variables in a calculated column, the scope of the variable is limited to the current row. Let's see this in action using the following code snippet as an example of DAX that works but could be made clearer and more elegant using variables.

```
Spares Vs. Labor Costs "Classic"=
SWITCH(
 TRUE(),
'Sales'[SpareParts] - 'Sales'[LaborCost] > 0, "Spares Higher",
'Sales'[SpareParts] - 'Sales'[LaborCost] < 0, "Spares Lower",
'Sales'[SpareParts] - 'Sales'[LaborCost] = 0, "Same"
)
```

The previous piece of DAX shows how a calculated column is used to apply a different text depending on the comparison of two values (spare parts and labor cost). The only slight drawback with this code is that it entails repeating a piece of code (the calculation of spare parts minus labor costs) three times.

Using a variable, the repeated code snippet can be moved to the variable. The variable is then used inside the SWITCH() function and so needs only appear once in the code. This column could be added to the Sales table.

```
Spares Vs. Labor Costs =
VAR c_SalesToSparesCost =
'Sales'[SpareParts] - 'Sales'[LaborCost]
RETURN

SWITCH(TRUE(),
c_SalesToSparesCost > 0, "Spares Higher",
c_SalesToSparesCost < 0, "Spares Lower",
c_SalesToSparesCost = 0, "Same"
)
```

The output from either of these functions (for 2020) gives the result that you can see in Figure 15-13.

Make	SpareParts	LaborCost	Spares Vs. Labor Costs
Alfa Romeo	£500.00	£500.00	Same
Alfa Romeo	£2,250.00	£1,500.00	Spares Higher
Alfa Romeo	£1,050.00	£2,360.00	Spares Lower
Aston Martin	£500.00	£500.00	Same
Aston Martin	£8,250.00	£5,150.00	Spares Higher
Aston Martin	£11,300.00	£20,825.00	Spares Lower
Austin	£1,500.00	£1,000.00	Spares Higher
Austin	£750.00	£1,360.00	Spares Lower
Bentley	£750.00	£1,490.00	Spares Lower
BMW	£1,500.00	£1,000.00	Spares Higher
Bugatti	£2,200.00	£2,000.00	Spares Higher
Total	**£100,725.00**	**£184,490.00**	

Figure 15-13. *Using variables in calculated columns*

To reiterate - and to accentuate the main point - any variable used inside a calculated column will, like any reference to a column from the same table, only be able to "see" data from the row that is being processed. So if you need to use data from another table, you will need to apply the RELATED() or RELATEDTABLE() functions.

Remember as well that a calculated column is only recalculated when the calculated column is created or the table refreshed. Using variables in the DAX for calculated columns does not affect this principle.

Filtering Using a Measure as the Comparison Element

All the measures that you developed using the CALCULATE() function in Chapter 8 have one thing in common. They all use a hard-coded value or a field reference as the comparison element (the value on the right of the equals sign inside the filter parameter). Attempting to use a measure as the comparison element gave an error message (also shown in Chapter 8).

Using a variable as the comparison element in a CALCULATE() filter allows you to get around this limitation. You can see this in the following DAX snippet where the measure Most Expensive Sale is used directly (and not inside a FILTER() function) to filter a CALCULATE() function. I have added this to the Sales table.

```
Costliest Car =
var c_ExpensiveCar = [Most Expensive Sale]

RETURN

CALCULATE(SUM(Sales[SalePrice]),
'Sales'[SalePrice] = c_ExpensiveCar)
```

You can see the output from this measure in Figure 15-14.

Make	Total Sales	Costliest Car
Alfa Romeo	£243,510	£25,000
Aston Martin	£5,102,755	£225,000
Austin	£64,500	£23,600
Bentley	£1,689,240	£189,500
BMW	£60,500	£33,500
Bugatti	£1,915,500	£365,000
Citroen	£101,080	£65,890
Delahaye	£107,000	£39,500
Delorean	£99,500	£99,500
Ferrari	£5,540,850	£395,000
Jaguar	£882,955	£99,500
Lagonda	£218,000	£156,500
Lamborghini	£1,727,150	£305,000
McLaren	£295,000	£295,000
Mercedes	£471,115	£62,500
Morgan	£18,500	£18,500
Noble	£206,900	£55,000
Peugeot	£21,045	£3,950
Porsche	£1,169,240	£89,500
Reliant	£1,900	£1,900
Rolls Royce	£1,637,000	£182,500
Trabant	£8,440	£2,500
Triumph	£396,270	£39,500
Total	**£21,977,950**	**£395,000**

Figure 15-14. *Using a measure attributed to a variable inside a CALCULATE()*
comparison

Simply placing the Most Expensive Sale measure inside the CALCULATE() function will *not* work. Using a variable to pass this measure as the comparison element used to filter a CALCULATE() function will solve the problem efficiently and effortlessly.

Commenting DAX

You may have started to appreciate that DAX can become complex. While this cannot always be helped, you can mitigate the effects by annotating your code with comments.

There are two ways to add comments to DAX:

- Comment out a single line or until the end of the line.

- Comment out a block of code that can be part of a line or multiple lines.

Commenting Lines

To comment out a line of code or to add a comment at the end of a line of code, you add two forward slashes where you want the comment to start.

This is a line that has been commented out:

```
// Here calculating the output after the RETURN
```

This is a line of DAX containing working code up until the comment starts

```
VAR c_FinalOutput // Here adding a comment after active code
```

Commenting Blocks of Code

Alternatively you can comment out a block of code of any length.

This is DAX containing an inline comment (note that this is not a complete DAX expression):

```
FILTER(Sales, Sales[CostPrice]
> /* Remember to update the number */ 50000)
,"Make", RELATED(Vehicle[Make])
,"Sale Price", Sales[SalePrice]
)
```

This is a line of DAX containing a comment block

```
Specific Countries =
/*
Using the IN function
To add a set of countries
*/
CALCULATE(SUM(Sales[SalePrice]), 'Geography'[CountryName] IN {"France",
"United States", "Belgium"})
```

You can see both of these approaches in the following code snippet that annotates a measure that you created previously.

```
Spares Vs. Labor Costs =
VAR c_SalesToSpatesCost = 'Sales'[SpareParts] - 'Sales'[LaborCost] //
Initial Variable
/*
A comment
Over
Several
Lines
*/
RETURN
// Here calculating the output after the RETURN
SWITCH(TRUE(),
c_SalesToSpatesCost > 0, "Spares Higher",
c_SalesToSpatesCost < 0, "Spares Lower",
c_SalesToSpatesCost = 0, "Same"
)
```

There are a couple of keyboard shortcuts that can help you when adding comments to DAX in Power BI Desktop. They are

- Crtl+KC will comment out any active line or selected lines

- Crtl+KU will remove comments from any active line or selected lines

Conclusion

In this chapter you discovered DAX variables. You saw how these can be used to simplify and clarify the DAX measures that you create.

You saw that variables can contain numbers, text calculations, and tables. You then learned how to create and apply variables in multiple ways to DAX measures and calculated columns. Finally you saw how to annotate your code by adding comments.

The measures added in this chapter can be found in the file PrestigeCarsDimensionalWithVariables.Pbix. This file is available to download alongside all the other sample files for the book from the Apress website.

The next extension to your DAX knowledge concerns DAX tables. These are what you will discover in the next chapter.

CHAPTER 16

Table Functions

One frequent DAX challenge that many users face is the need to deliver highly specific, often comparative, data. You may be required, say, to look at how data overlaps when different criteria are applied. Maybe you will need to look at disparate subsets of data and compare and contrast (or even summarize) them.

To solve problems like these (and to provide solutions to many other calculation challenges as well), you will need to use *table functions* in your DAX.

Table functions allow you to create *virtual datasets* that provide a subset of data that you can tailor to your own requirements. You create these virtual datasets to contain only the rows and columns that you want. Then you use their output in a myriad of ways to deliver the result that you are looking for.

While table functions are incredibly powerful, they can get very complex both to understand and to write. So, in this chapter, I am going to focus on the core aspects of *how* you write table functions. As table functions can be used in many different ways, it is impossible here to show more than a small cross-section of the myriad ways they can be used to solve real-world DAX challenges.

This chapter is going to make heavy use of variables. So if you have not already read Chapter 15 (or if you need a refresher on variables first), it would be a good idea to read (or skim through) the previous chapter first.

This chapter will use the Power BI Desktop file PrestigeCarsDimensionalWith Variables. pbix as the source of data in the examples that I will show. So you will need to download this file from the Apress website if you want to practice the exercises in this chapter.

Table Variables in Table Functions

Table functions nearly always rely on the use of table variables. This is not only because using variables makes DAX easier to write, test, and understand. In the case of table functions, it is because using variables to store the output from a table function is nearly always the only practical way to write the DAX.

© Adam Aspin 2023
A. Aspin, *Pro DAX and Data Modeling in Power BI*, https://doi.org/10.1007/978-1-4842-8995-2_16

As a first example, suppose that you want to find the value of the top five selling models sold. The measure that returns this figure can then be sliced and diced.

The upcoming DAX makes use of the TOPN() DAX function. This function returns a specified number of rows from a designated table based on a defined sort order. To do this it needs four parameters:

- The number of rows to return

- A table containing the data from which you want to return records

- The field in this table that will be used as a basis for ordering the rows

- The sort order to apply

One way of delivering this metric is to write the following DAX (added to the Sales table in this example):

```
Top Sales =
VAR t_SalesByProduct =
SUMMARIZECOLUMNS(
 'Vehicle'[Model]
,"Total Sales", SUM(Sales[SalePrice])
)

VAR t_BestSellingFiveModels =
TOPN(5, t_SalesByProduct, [Total Sales], DESC)

VAR c_TopSellersAmount =
CALCULATE(SUM(Sales[SalePrice]), t_BestSellingFiveModels)

RETURN
C_TopSellersAmount
```

Figure 16-1 illustrates how this measure can be used. I have also added the unfiltered output from this measure to make the point that the result is a standard measure that can subsequently be sliced filtered (or not as is the case for the lower figure) in any way you choose.

Figure 16-1. *Using a table function to output the cumulative top sales measure*

This DAX snippet introduces a core table function. This is SUMMARIZECOLUMNS().
What this function does is to create a virtual table. In this particular example the table is
the list of models sold and the total sales per model (if you look ahead to Figure 16-2 you
can see a sample of this output).

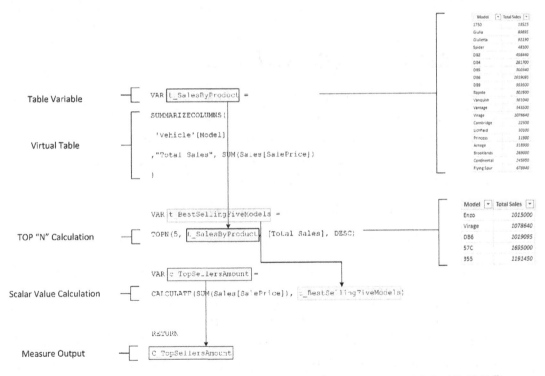

Figure 16-2. *Understanding table functions and SUMMARIZECOLUMNS()*

So what the entire DAX measure Top Sales does is

1. Create a virtual table of sales by model. This virtual table is stored in the variable SalesByProduct.

2. Uses the table variable SalesByProduct as the input to a TOPN() function that also specifies as its second and third parameters the number of elements to extract and the field to use to order the data by. The output from the TOPN() function - which is also a table variable - is passed to the variable BestSellingFiveModels.

3. The table variable BestSellingFiveModels (that contains the list of the five top selling models) is used as a *filter* inside a CALCULATE() function that returns a single value. This value is the sum of the sale price for the top five models sold.

To give you a more visual way of understanding how this code works, you can take a look at Figure 16-2.

The SUMMARIZECOLUMNS Function

SUMMARIZECOLUMNS() is not only the first table function that you could find yourself using, it is also one of the most powerful. So it is worth taking a deeper look at this specific function before going any further with other table functions.

SUMMARIZECOLUMNS() is designed to assemble, filter, and aggregate source data into a single tabular subset of data. It is based around

1. A list of fields from the data model that you want to use in the virtual table that the function creates

2. Any filters that you need to apply

3. Any aggregations you want to add in the output

As SUMMARIZECOLUMNS() is such a supple yet powerful function, it is worth taking a look at a few of the various ways that you can apply it create virtual tables. Also it is worth noting from the start that the second and third elements in the preceding list are optional – you can see this (along with other variations on the theme of using SUMMARIZECOLUMNS()) in the following sections.

You need to be aware from the start that SUMMARIZECOLUMNS() can take many parameters. However you need to ensure that the parameters are "grouped" in that columns must come first, filters must come second, and aggregations last - or the function will not work.

Column List Without Filters or Aggregations

The simplest use of SUMMARIZECOLUMNS() is to return a list of attributes. The output can be from any table in the data model. All you have to do is to add a comma-separated list of fields from the data model as parameters to the SUMMARIZECOLUMNS() function like this.

```
SUMMARIZECOLUMNS
(
  'Client'[ClientName]
,'Vehicle'[Make]
,'Vehicle'[Model]
)
```

Figure 16-3 shows the table returned from this DAX snippet. Here the DAX is used as the basis for a new table (created by clicking on any table in the Fields pane and then selecting New table in the Table tools ribbon).

Model ▼	ClientName ▼	Make ▼
308	Alex McWhirter	Ferrari
308	Alexei Tolstoi	Ferrari
308	Alicia Almodovar	Ferrari
308	Andrea Tarbuck	Ferrari
308	Andy Cheshire	Ferrari
308	Antonio Maura	Ferrari
308	Autos Sportivos	Ferrari
308	Beltway Prestige Driving	Ferrari
308	Birmingham Executive Prestige Vehicles	Ferrari
308	Bling Bling S.A.	Ferrari
308	Bling Motors	Ferrari
308	Boris Spry	Ferrari
308	Bravissima!	Ferrari
308	Capots Reluisants S.A.	Ferrari
308	Casseroles Chromes	Ferrari
308	Clubbing Cars	Ferrari

Figure 16-3. *Basic output from the SUMMARIZECOLUMNS() function*

Even when returning text (or attributes if you prefer), SUMMARIZECOLUMNS() is aggregating the data. So the output contains one row for each of the available combinations of the chosen fields.

SUMMARIZECOLUMNS() can access the entire data model and consequently does not need the RELATED() or RELATEDTABLE() functions. However you *have* to add fully qualified field names - that is, table and field names - so that DAX can identify the correct field to use.

Column List with Filters

Once you have selected a set of columns to return in the output from SUMMARIZECOLUMNS(), you will probably want to filter the dataset that it returns.

You do this by adding one or more FILTER() elements after the column names as a further parameter for the SUMMARIZECOLUMNS() function. You can see this in the following short piece of DAX.

```
SUMMARIZECOLUMNS
(
  'Client'[ClientName]
 ,'Vehicle'[Make]
 ,'Vehicle'[Model]
 ,'Vehicle'[Color]
 ,FILTER(VALUES('Vehicle'[Color]), 'Vehicle'[Color] = "Blue")
)
```

What you have done here is to

1. Add a FILTER() function after the list of columns to group and output.

2. Specify the field to filter on as the first parameter of the FILTER() function wrapped inside a VALUES() function.

Placing the field to filter on inside VALUES() is required when filtering data output from SUMMARIZECOLUMNS(). This is to ensure that only a *list of unique records constitutes the table* that FILTER() iterates over when filtering the dataset.

Figure 16-4 shows the table returned from this DAX snippet.

Model	ClientName	Make	Color
DB2	Alex McWhirter	Aston Martin	Blue
DB2	Alexei Tolstoi	Aston Martin	Blue
DB2	Alicia Almodovar	Aston Martin	Blue
DB2	Andrea Tarbuck	Aston Martin	Blue
DB2	Andy Cheshire	Aston Martin	Blue
DB2	Antonio Maura	Aston Martin	Blue
DB2	Autos Sportivos	Aston Martin	Blue
DB2	Beltway Prestige Driving	Aston Martin	Blue
DB2	Birmingham Executive Prestige Vehicles	Aston Martin	Blue
DB2	Bling Bling S.A.	Aston Martin	Blue
DB2	Bling Motors	Aston Martin	Blue
DB2	Boris Spry	Aston Martin	Blue

Figure 16-4. Adding a FILTER() function to SUMMARIZECOLUMNS()

You do not have to add the column that you are filtering on to the output. However this can be useful while testing and debugging, for example, as I have done here.

You can, if you need, add multiple FILTER() elements. Doing this will create AND filtering logic that reduces the number of rows in the virtual table.

Adding a filter inside SUMMARIZECOLUMNS() is not compulsory. However if you add a filter, it must be placed *after* all the columns that you want the output table to contain.

Column List with Aggregated Values

The final elements that you can add to a SUMMARIZECOLUMNS() function are aggregations. You can aggregate as many values from the dataset as you want provided that

- Aggregations follow filter elements (or if there are no filter elements, the column list)

- Aggregations always have an output column name defined, in quotes separated by a comma from the formula that is to be applied

This is much simpler than it probably sounds - as you can see from the following DAX snippet:

```
SUMMARIZECOLUMNS
(
  'Client'[ClientName]
 ,'Vehicle'[Make]
 ,'Vehicle'[Model]
,"Total Cost", SUM(Sales[CostPrice])
,"Total Sales", SUM(Sales[SalePrice])
)
```

Figure 16-5 shows the table returned from this DAX snippet.

Model	ClientName	Make	Total Cost	Total Sales
DB6	Vive La Vitesse	Aston Martin	66314	90500
Vanquish	Laurent Saint Yves	Aston Martin	46636	65500
Rapide	Screamin' Wheels	Aston Martin	44880	55000
DB9	M. Pierre Dubois	Aston Martin	42804	56500
DB6	Alicia Almodovar	Aston Martin	37495	45950
Virage	ImpressTheNeighbours.Com	Aston Martin	85600	100000
DB9	Francoise LeBrun	Aston Martin	34092	45000
DB9	Stan Collywobble	Aston Martin	36480	45600
DB5	Prestissimo!	Aston Martin	33185	45000
Vanquish	Magic Motors	Aston Martin	41668	55000

Figure 16-5. Adding aggregations to SUMMARIZECOLUMNS()

Something else to be aware of is that there is no need to sort the data in a virtual table. Sorting can be applied in visuals if required.

Column List, Filter, and Aggregation

Finally you need to see how all the three component parameter groups (fields to group and output, filters, and aggregated values) of a SUMMARIZECOLUMNS() function can work together. Building on the previous examples, you can see column selection, filtering, and aggregation used together in the following DAX snippet.

```
SUMMARIZECOLUMNS
(
  'Client'[ClientName]
,'Vehicle'[Make]
,'Vehicle'[Model]
,FILTER(VALUES('Vehicle'[Color]), 'Vehicle'[Color] = "Blue")
,"Total Cost", SUM(Sales[CostPrice])
,"Total Sales", SUM(Sales[SalePrice])
)
```

Figure 16-6 shows the table returned from this DAX snippet. You can see that the output

- Is aggregated by Model, Client, and Make

- Is filtered by blue cars (you can test this by adding 'Vehicle'[Color] to the list of fields if you wish)

- Aggregates the totals for cost price and sale price

Model ▼	ClientName ▼	Make ▼	Total Cost ▼	Total Sales ▼
404	London Executive Prestige Vehicles	Peugeot	2996	3500
Torpedo	Autos Sportivos	Citroen	56402	65890
924	Alicia Almodovar	Porsche	10700	12500
Giulietta	Prestissimo!	Alfa Romeo	5564	6500
Isetta	Silver HubCaps	BMW	4708	5500
1750	Vive La Vitesse	Alfa Romeo	8517	9950
944	Screamin' Wheels	Porsche	6420	7500
Virage	ImpressTheNeighbours.Com	Aston Martin	85600	100000
Giulietta	Kieran O'Harris	Alfa Romeo	15836	18500
M14	Melissa Bertrand	Noble	4708	5500
500	Mary Blackhouse	Trabant	2140	2500

Figure 16-6. *SUMMARIZECOLUMNS() to group, filter, and aggregate data*

As this is a more complex and complete application of the SUMMARIZECOLUMNS() function, Figure 16-7 shows the way that this DAX snippet works more graphically.

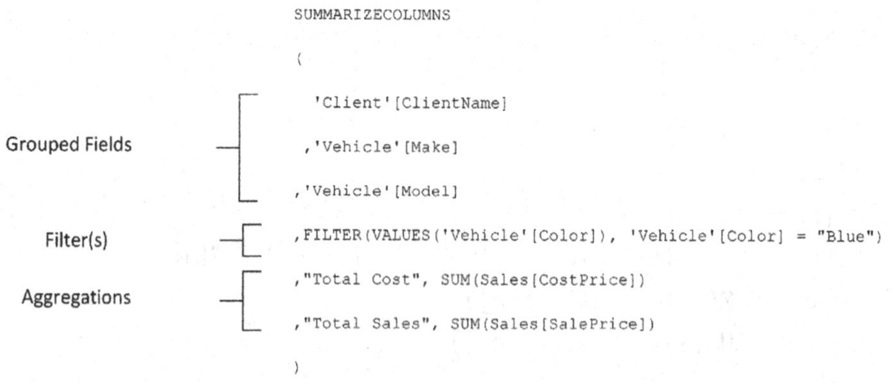

```
SUMMARIZECOLUMNS

(
                          'Client'[ClientName]

Grouped Fields            ,'Vehicle'[Make]

                          ,'Vehicle'[Model]

Filter(s)                 ,FILTER(VALUES('Vehicle'[Color]), 'Vehicle'[Color] = "Blue")

                          ,"Total Cost", SUM(Sales[CostPrice])
Aggregations
                          ,"Total Sales", SUM(Sales[SalePrice])

)
```

Figure 16-7. *The anatomy of a complex SUMMARIZECOLUMNS() function*

Adding Columns to the Output from Table Functions

Table functions are not limited to generating virtual tables to use in measures. They can also carry out a whole series of operations on existing tables. One classic requirement when using table functions is to add columns to existing tables. You can do this using the ADDCOLUMNS() function.

As the function name describes what it does, let's see it in action. In this example the requirement is to order sales by model of car. One way of doing this is as follows.

```
Ranked Sales By Model =
VAR t_SalesByProduct =
SUMMARIZECOLUMNS(
 'Vehicle'[Model]
,"Total Sales", SUM(Sales[SalePrice])
)

VAR t_OrderColumn =
ADDCOLUMNS(t_SalesByProduct, "OrderOfSales", RANKX(t_SalesByProduct, [Total
Sales]))
RETURN
t_OrderColumn
```

What this DAX does is

1. An initial variable - t_SalesByProduct - holds the table containing the aggregate sales per model.

2. t_SalesByProduct is used by the ADDCOLUMNS() function as the source table that this function extends with an additional column.

3. The additional column is defined by a RANKX() function that returns the number in the sale ranking for each row in the t_SalesByProduct table variable.

We are not actually using the output from this DAX yet. I am leaving that until the next section. However for completeness, Figure 16-8 nonetheless shows the table that is generated by the previous DAX snippet. Note that this output has not been sorted by order of sales.

Model		Total Sales		OrderOfSales	
1750		13525		75	
Giulia		89695		50	
Giulietta		92190		48	
Spider		48100		62	
DB2		456440		15	
DB4		281700		28	
DB5		300340		25	
DB6		1019095		4	
DB9		959500		6	
Rapide		302500		24	

Figure 16-8. *Using the ADDCOLUMNS() table function*

Filtering Table Function Output

After a reasonably in-depth look at SUMMARIZECOLUMNS(), it is time to move back to the overview of other available table functions. The next topic on the agenda is to filter table function output. One way of doing this is to take the output from one table function and then apply a filter to this in its turn.

One challenge that requires table function output filtering is to display the total sales for the next ten sales by model that follow the top ten sales by model. The following DAX code snippet does exactly this (and it does this using the DAX that you created in the previous section). I suggest adding this measure to the SalesInfo table.

```
Next 10 Models =
VAR t_SalesByProduct =
SUMMARIZECOLUMNS(
 'Vehicle'[Model]
,"Total Sales", SUM(Sales[SalePrice])
)

VAR t_OrderColumn =
ADDCOLUMNS(t_SalesByProduct, "OrderOfSales", RANKX(t_SalesByProduct, [Total
Sales]))
```

```
VAR t_NextTen =
FILTER(t_OrderColumn,
[OrderOfSales] > 10 && [OrderOfSales] <= 20)

RETURN
CALCULATE(SUM(Sales[SalePrice]), t_NextTen)
```

Figure 16-9 shows the output from the measure that uses table function filtering.

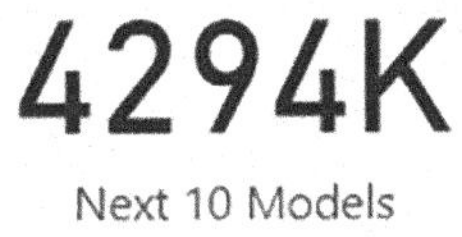

4294K

Next 10 Models

Figure 16-9. *Filtering the output from a table function*

The way that this piece of code works is as follows:

1. An initial variable - t_SalesByProduct - holds the table containing the aggregate sales per model.

2. A second variable - t_OrderColumn - extends this table with a column that adds the sales ranking.

3. The final variable - t_NextTen - applies a FILTER() function to isolate the records for sales ranked between 11 and 20 by value.

4. The measure outputs the total of sales using the final (t_NextTen) table variable as the filter in the CALCULATE() function.

Figure 16-10 tries to explain this process more graphically.

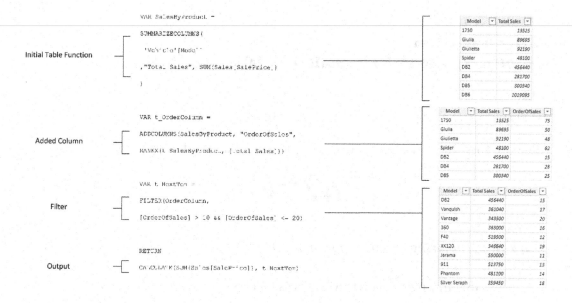

Figure 16-10. *Anatomy of a filtered table function*

Advanced Filtering Using CALCULATETABLE( )

Just as you have CALCULATE() when filtering (or unfiltering) measures, you have
CALCULATETABLE() to filter - or extend and adapt filters - on DAX tables.

One classic way to use CALCULATETABLE() among all its plethora of possibilities is
to use it to filter the output from an existing table variable. As an example imagine that
you want to see the top two makes by sales and also define a date range for this. The
following DAX shows one way of achieving this using CALCULATETABLE().

```
Top Recent Sellers =

VAR t_TopSellersTable =
TOPN(
2,
SUMMARIZECOLUMNS
(
Vehicle[Make]
,"Total Sales", SUM(Sales[SalePrice])
),
[Total Sales]
```

```
,DESC
)
VAR t_DateFilteredTopSellers =

CALCULATETABLE(
SUMMARIZECOLUMNS(
CLient[ClientName]
,Geography[Town]
,Vehicle[Make]
,"Sales" , SUM(Sales[SalePrice])
)
,t_TopSellersTable
,DATESBETWEEN('Date'[DateKey], DATEVALUE("1/1/2022"),
DATEVALUE("1/31/2022"))
)

RETURN
T_DateFilteredTopSellers
```

Figure 16-11 shows the table created from this DAX snippet.

ClientName	Town	Make	Sales
King Leer Cars	Newcastle	Aston Martin	56850
Clubbing Cars	Manchester	Aston Martin	56950
Honest Pete Motors	Stoke	Aston Martin	42500
Silver HubCaps	Manchester	Aston Martin	62500
Screamin' Wheels	Washington	Aston Martin	65450

Figure 16-11. *Using CALCULATETABLE()*

The way that this piece of DAX works is as follows:

1. An initial variable – SalesByProduct - holds the table containing the top two sales per make. This will be used later in the code as a filter.

2. A table variable DateFilteredTopSellers aggregates a table of clients, makes, town, and sales.

3. CALCULATETABLE() is used to modify the way that the output from SUMMARIZECOLUMNS() is filtered. In this example two filter modifiers are added as the second and third parameters to CALCULATETABLE(). These modifiers

    a. Use the TopSellersTable as a filter

    b. Apply a date range filter

As this is a complex piece of DAX function, Figure 16-12 shows the way that this DAX snippet works more graphically.

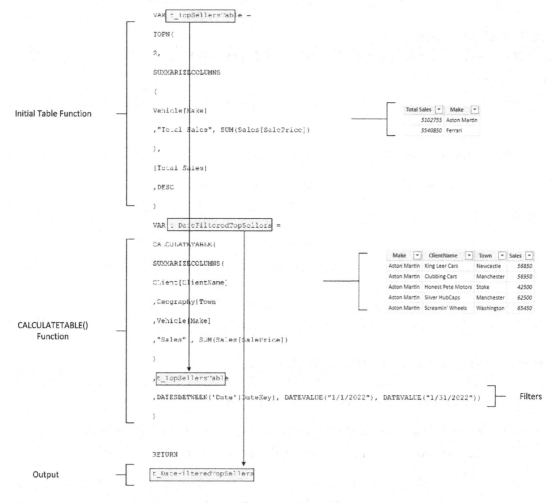

Figure 16-12. CALCULATETABLE() explained

CALCULATETABLE() is very similar to CALCULATE() in that

- The first parameter is the output. In the case of CALCULATETABLE() this (perhaps inevitably) has to be a table and not a scalar value.

- All other parameters are filters or modifiers.

In this example I have hard-coded the date range. In practice you could want to specify the last month or the last few months so that you output a table that is updated each time a Power BI report is refreshed. If you need a quick revision of time intelligence functions needed to do this, then you can find the necessary information in Chapter 14.

Removing Columns

Just as you can add columns to the output from a table function, you can also remove them. To illustrate this I am going to take the DAX that you saw previously when calculating the top 10 sales, and clean it up a little.

The DAX that has extended a previous snippet (and that I have added to the Sales table) is

```
Best 10 Sales =
VAR t_SalesByProduct =
SUMMARIZECOLUMNS(
 'Vehicle'[Model]
,"Total Sales", SUM(Sales[SalePrice])
)

VAR BestSellingModels =
TOPN(10, t_SalesByProduct, [Total Sales], DESC)

VAR t_Best10 =
SELECTCOLUMNS(BestSellingModels, "Model", 'Vehicle'[Model])

RETURN
CALCULATE(SUM(Sales[SalePrice]), t_Best10)
```

The output from the BestSellingModels and Best10 table variables are shown alongside the measure used in a card in Figure 16-13.

Model	
Flying Spur	
F50	
Diabolo	
DB9	
Testarossa	
Virage	
355	
DB6	
Enzo	
57C	

Model	Total Sales
Flying Spur	678940
F50	760950
Diabolo	905000
DB9	959500
Testarossa	870000
Virage	1078640
355	1191450
DB6	1019095
Enzo	1015000
57C	1695000

10M

Best 10 Sales

Measure Output   Best10   BestSellingModels

Figure 16-13. *Using SELECTCOLUMNS() to reduce DAX table output*

This piece of DAX shows you how SELECTCOLUMNS() can be used to remove columns from a table function's output.

SELECTCOLUMNS

As befits a product that lives and breathes aggregated data, we have spent all of this chapter so far looking at table functions that aggregate data. However, DAX table functions can also output tables without any aggregation. This can be useful when you are looking to return the number of records in a custom dataset where you need to look at the detail not the aggregation level, for instance.

Simple SELECTCOLUMNS( )

At its simplest, SELECTCOLUMNS() lets you

- Specify a table to choose output columns from
- Specify the column names you want to use in the output

You can see this in the following short DAX snippet.

```
SELECTCOLUMNS
(
Vehicle
,"Make", [Make]
,"Model", [Model]
,"Color", [Color]
)
```

All this does is to create a subset of data from a table. In itself this is probably not terribly interesting. However, it is a first step to more complex table functions, so let's progress to something a little more complex using SELECTCOLUMNS()

SELECTCOLUMNS( ) Across Multiple Tables

You can also use SELECTCOLUMNS() to return a dataset consisting of fields from different tables. The tables will, for the moment at least, have to be joined in a coherent data model.

This means that you could, for instance, write DAX like this to list data from the Sales, Vehicle, and Geography tables.

```
SELECTCOLUMNS(
Sales
,"Make and Model", RELATED(Vehicle[Make]) & " - " &
                   RELATED(Vehicle[Model])
,"Town", RELATED(Geography[Town])
,"Gross Margin", Sales[SalePrice] - Sales[CostPrice]
)
```

As it probably helps to map output to code, you can see the table produced by this DAX in Figure 16-14 (even if you never use the output directly).

Make and Model	Town	Gross Margin
Alfa Romeo - 1750	Lausanne	658
Alfa Romeo - 1750	Marseille	1433
Alfa Romeo - Giulia	Brussels	1932
Alfa Romeo - Giulia	London	8162
Alfa Romeo - Giulia	Manchester	5040
Alfa Romeo - Giulia	Marseille	857
Alfa Romeo - Giulia	Stuttgart	2016
Alfa Romeo - Giulia	Washington	4200
Alfa Romeo - Giulietta	Barcelona	1548
Alfa Romeo - Giulietta	Berlin	3142
Alfa Romeo - Giulietta	Liverpool	2664
Alfa Romeo - Giulietta	London	1932
Alfa Romeo - Giulietta	Los Angeles	9182
Alfa Romeo - Giulietta	Milan	936
Alfa Romeo - Spider	Brussels	2300
Alfa Romeo - Spider	London	2300
Total		**5121287**

Figure 16-14. SELECTCOLUMNS() across multiple tables

This particular DAX snippet has kept the basic principles that you discovered in the previous section. That is, you start by specifying a source table as the first parameter of the SELECTCOLUMNS() function and then continue with the columns required in the output as the remaining parameters of the function.

The difference here is that any columns that are not sourced from the table referenced in the first parameter have to be "joined" using the RELATED() function. This means that (unless you are applying some of the more advanced techniques that are explained in the next chapter) you need to choose as the first parameter a table that is linked in the data model to other tables in a way that allows RELATED() to be used. This is explained in Chapter 5. In this example the fact table Sales is the "many" side of relationships to the Geography and Vehicle tables so is the best choice for the table to use as the core of the SELECTCOLUMNS() function.

Also - and as you can see in the preceding DAX snippet - you can add calculations to the output from SELECTCOLUMNS() just as you can in a calculated column. So if the fields to use in a calculation are in the table used as the first parameter, you can reference them directly. If they are in other tables that can be accessed in the data model, you can reference them using RELATED().

Filtering SELECTCOLUMNS( )

Now let's put it all together and really use SELECTCOLUMNS() in a practical way. The objective is to isolate a virtual table that lists individual car sales of over £50,000 so that you can use the resulting measure sliced and diced in different ways.

One way to do this (using SELECTCOLUMNS()) is to create a measure using DAX like this.

```
Some Expensive Sales =
VAR t_ExpensiveSales =
SELECTCOLUMNS
(
FILTER(Sales, Sales[CostPrice] > 50000)
,"Make", RELATED(Vehicle[Make])
,"Sale Price", Sales[SalePrice]
)

RETURN
COUNTROWS(t_ExpensiveSales)
```

You can see an example of the output in Figure 16-15. As the table function is now part of a measure that returns a single value, its output can be filtered and sliced.

MonthAbbrAndYearAbbr	Some Expensive Sales
Jan 19	1
Apr 19	1
May 19	2
Oct 19	1
Dec 19	1
Jan 20	2
Mar 20	1
Apr 20	2
Jun 20	1
Jul 20	4
Aug 20	6
Sep 20	2
Oct 20	2
Dec 20	5
Total	**31**

Figure 16-15. *Adding a FILTER() function when using SELECTCOLUMNS()-*

As is so often the case when writing DAX measures, this is not the only (or, arguably, the best) way to resolve the challenge. However it does show you how table functions - and specifically SELECTCOLUMNS() – can be applied in a practical way.

INTERSECT

One of the core reasons for using table functions is to isolate datasets that allow for some form of comparison. As an example of this, suppose that you want to display models that are neither the top sellers nor the worst selling cars. In fact you want to see a specified number of models that are "middle-sellers" - neither particularly good nor particularly bad.

One way to do this is to use a pair of datasets that are created using table functions. In this case, let's see the code and its output first and explain it afterwards. The DAX to do this is

```
Best and Worst Seller Intersection =
VAR t_SalesByProduct =
SUMMARIZECOLUMNS(
 'Vehicle'[Model]
```

```
,"Total Sales", SUM(Sales[SalePrice])
)

VAR c_NumberOfRecords = COUNTROWS(SalesByProduct)

VAR t_BestSellingModels =
TOPN(c_NumberOfRecords/2 +2, t_SalesByProduct, [Total Sales], DESC)

VAR t_WorstSellingModels =
TOPN(c_NumberOfRecords/2 +2, t_SalesByProduct, [Total Sales], ASC)

VAR t_BestSellers =
INTERSECT(t_BestSellingModels, t_WorstSellingModels)

VAR s_BestAndWorst =
CONCATENATEX(t_BestSellers, 'Vehicle'[Model], ", ")

RETURN
s_BestAndWorst
```

Figure 16-16 illustrates how the Best and Worst Seller Intersection measure can be used in a card visual.

280SL, 959, Phantom, Torpedo, Vantage

Best and Worst Seller Intersection

Figure 16-16. Output from the INTERSECT() function using CONCATENATEX()

This piece of DAX works as follows:

1. An initial variable – SalesByProduct - holds the table containing the aggregate sales per model.

2. A DAX variable uses the COUNTROWS() function to work out how many rows there are in this virtual table.

3. A second table variable, BestSellingModels, isolates a dataset containing the top 50 percent of records (calculated by dividing the row count by two). A fixed number then defines the overlap you want from the top records.

4. A third table variable, WorstSellingModels, isolates a dataset containing the top 50 percent of records (calculated by dividing the row count by two). A fixed number then defines the overlap you want from the bottom records.

5. The INTERSECT() function isolates any records that are common to the two datasets containing, respectively, the top 50% plus two and bottom 50% plus two records.

6. The CONCATENATEX() function. This extracts the model records from the t_BestSellers virtual table and joins all the elements into a single value that can be returned as the scalar output from the measure.

As you can see here, DAX variables lend themselves to reuse once again. Indeed, in this short piece of DAX, you are applying the initial table variable SalesByProduct three times and the NumberOfRecords scalar variable twice.

The main function that I want to explain here is the INTERSECT() function. INTERSECT() is a table function that takes two datasets and returns a table of all overlapping elements. Put graphically, it does as you can see in Figure 16-17.

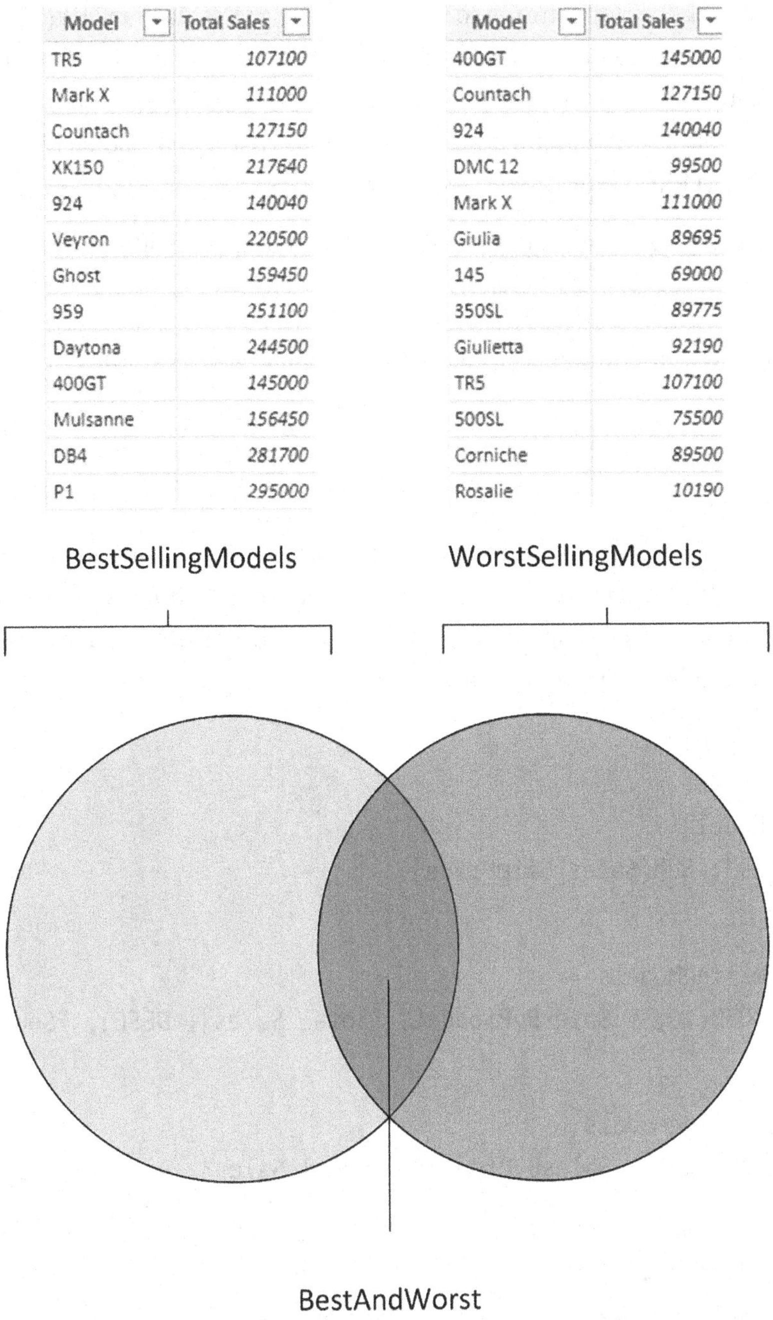

Model	Total Sales
TR5	107100
Mark X	111000
Countach	127150
XK150	217640
924	140040
Veyron	220500
Ghost	159450
959	251100
Daytona	244500
400GT	145000
Mulsanne	156450
DB4	281700
P1	295000

Model	Total Sales
400GT	145000
Countach	127150
924	140040
DMC 12	99500
Mark X	111000
Giulia	89695
145	69000
350SL	89775
Giulietta	92190
TR5	107100
500SL	75500
Corniche	89500
Rosalie	10190

BestSellingModels   WorstSellingModels

BestAndWorst

Figure 16-17. *Outputting table variables using the INTERSECT() function*

I am also taking the opportunity to introduce The CONCATENATEX() function. This is a way of taking the data contained in a column for all the rows in a table (the INTERSECT() function in this example) and making multiple cells into a single scalar value. As the output is no longer a table, it can now be used in a visuals directly.

CONCATENATEX() takes three parameters:

- The table whose records contain a column to be concatenated

- The actual column to concatenate

- The separator character to use (the comma in this example)

UNION

The UNION() function is the second of the three core table functions that can be used to "compile" datasets based on existing output from other table functions.

To see this in practice, imagine that you need a table of the best and worst selling cars. The following DAX can create this.

```
VAR t_SalesByProduct =
SUMMARIZECOLUMNS(
 'Vehicle'[Model]
,"Total Sales", SUM(Sales[SalePrice])
)

VAR t_BestSellingModels =
ADDCOLUMNS(TOPN(10, t_SalesByProduct, [Total Sales], DESC), "Source", "Best
Sellers")

VAR t_WorstSellingModels =
ADDCOLUMNS(TOPN(10, t_SalesByProduct, [Total Sales], ASC), "Source", "Worst
Sellers")

VAR t_BestAndWorst =
UNION(t_BestSellingModels, t_WorstSellingModels)

RETURN
t_BestAndWorst
```

Figure 16-18 shows this DAX table function used as the basis for a new table. This was created by

1. Clicking on any table in the Fields pane (or switch to Data view)

2. In the table tools menu clicking New table

3. Adding the table name

4. Adding the DAX and confirming

Model	Source	Total Sales
355	Best Sellers	1191450
57C	Best Sellers	1695000
DB6	Best Sellers	1019095
DB9	Best Sellers	959500
Diabolo	Best Sellers	905000
Enzo	Best Sellers	1015000
F50	Best Sellers	760950
Flying Spur	Best Sellers	678940
Testarossa	Best Sellers	870000
Virage	Best Sellers	1078640
175	Worst Sellers	12500
203	Worst Sellers	3200
205	Worst Sellers	4900
500	Worst Sellers	3650
600	Worst Sellers	4790
Isetta	Worst Sellers	5500
M14	Worst Sellers	5500
Princess	Worst Sellers	11900
Robin	Worst Sellers	1900
Rosalie	Worst Sellers	10190
Total		**10237605**

Figure 16-18. *Output from the UNION() function*

Once the table is created, you can use the fields it contains in any visual. However, in this particular example, it is best suited to a table visual.

Creating tables from table functions that you then use in dashboards is, in practice, fairly rare. Yet it is worth knowing that it can be done. However, you need to be aware of the limitations that come with this technique:

- The content of the table is static (sounaffected by filters and slicers in the data model). This holds true even if you join the new DAX-driven table to other tables.

- The table is only refreshed when the underlying tables that contain the source data it uses are refreshed.

- You cannot refresh the table directly.

EXCEPT

EXCEPT() is the last of the three dataset amalgamation table functions in DAX. It works on the same underlying principle as INTERSECT() and UNION(); only this function returns the rows of left-side table which do *not* appear in right-side table.

Take a look at the following DAX that isolates a dataset containing models that sold well that are not red.

```
Not Good or Red Sales =
VAR t_SalesByProduct =
SUMMARIZECOLUMNS(
 'Vehicle'[Model]
,'Vehicle'[Color]
,"Total Model Color Sales", SUM(Sales[SalePrice])
)

VAR t_GoodSales =
FILTER(t_SalesByProduct, [Total Sales] > 250000)

VAR t_RedCars =
FILTER(t_SalesByProduct, [Color] = "Red")

VAR t_GoodNotRed =
EXCEPT(t_GoodSales, t_RedCars)

RETURN
CALCULATE(SUMX(t_GoodNotRed, [Total Model Color Sales]))
```

Note   As is the case with the INTERSECT() and UNION() functions, it is *vital* to ensure that the two tables used for amalgamation, exclusion, or compilation have identical structures. This means having the same number of columns in the same order. Columns do not have to have the same names as the column names from the first table will be used in the output.

There are certainly easier and better ways of producing this output. However, the aim here is to introduce the EXCEPT() function as simply as possible. What this example does is

1. Aggregates sales by color and model in the SalesByProduct table variable.

2. Creates a table variable - t_GoodSales - that filters the t_SalesByProduct table variable to exclude aggregate sales less than £250,000 per model and color combination.

3. Creates a table variable - t_RedCars - that filters the t_SalesByProduct table to exclude all non-red cars.

4. Applies the EXCEPT() function to output all the rows from the t_GoodSales table variable that do not appear in the t_RedCars table variable. This is placed in the t_GoodNotRed variable.

5. Calculates the total of the sales in the t_GoodNotRed table variable. This requires an aggregation function as the output from a table variable is produced by the EXCEPT() function.

Figure 16-19 shows a card visual displaying the output for 2022.

£1,189,500

Not Good or Red Sales

Figure 16-19. Output from the EXCEPT() function

This output can be filtered or sliced by other attributes (but only *either* by color *or* model as both are included in the SalesByProduct table variable).

Figure 16-20 shows the process more graphically.

Model	Color	Total Model Color Sales
DB6	Black	375030
Virage	Black	292090
57C	Black	295000
355	Black	500450
Enzo	Black	620000
Mondial	Black	257950
Testarossa	Black	445000
Diabolo	Black	490000
57C	Blue	355000
F50	Blue	255950
Enzo	British Racing Green	395000
Jarama	Green	305000
Phantom	Green	302100
57C	Red	710000
F40	Dark Purple	269500
DB9	Silver	269900
57C	Silver	335000
F50	Silver	310000
P1	Silver	295000

Model	Color	Total Model Color Sales
Giulia	Red	8695
DB5	Red	45000
DB6	Red	45500
DB9	Red	79500
Rapide	Red	61500
Virage	Red	171400
Princess	Red	2500
Flying Spur	Red	55450
Mulsanne	Red	56950
57C	Red	710000
Veyron	Red	220500

GoodSales   RedCars

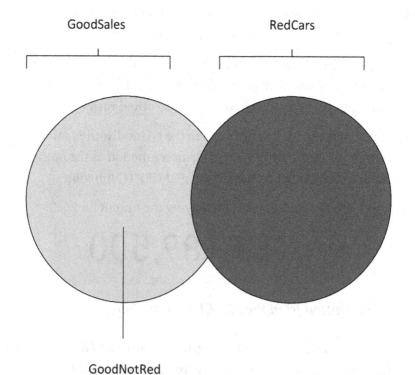

GoodNotRed

Figure 16-20. The EXCEPT() function explained

412

CROSSJOIN( )

Sometimes you need to apply OR selections to a DAX table when the criteria are from different fields. Say, for instance, that you want either a specific color or a specific make. Now, whereas providing OR selections from a single column is easy, applying alternative choices from different columns is a little more challenging.

This is one set of circumstances when the CROSSJOIN() function can come in useful. The following DAX code uses CROSSJOIN() to create a table of all the possible choices from inside two different columns. This table can then be used for an OR filter.

Admittedly, you saw this function briefly in Chapter 12. However the explanation was somewhat superficial when you first met this function so as not to distract from other points of focus. Now you can learn about CROSSJOIN() in more detail. Take a look at the following piece of DAX.

```
Aston Martin Or France =
VAR t_Multicriteria =

CROSSJOIN
(
 ALL(Geography[CountryName])
,ALL(Vehicle[Make])
)

VAR t_SelectedCriteria =

FILTER(
t_MultiCriteria
,
OR(
 Geography[CountryName] = "France"
,Vehicle[Make] = "Aston Martin"
)
)

VAR t_OrOutput =
```

```
SUMMARIZECOLUMNS
(
 Vehicle[Make]
,Vehicle[Model]
,Geography[CountryName]

,t_SelectedCriteria

,"Total Sales", SUM(Sales[SalePrice])
)

RETURN
SUMX(t_OrOutput, [Total Sales])
```

Figure 16-21 shows the result in a card visual for Aston Martin DB6 for 2020.

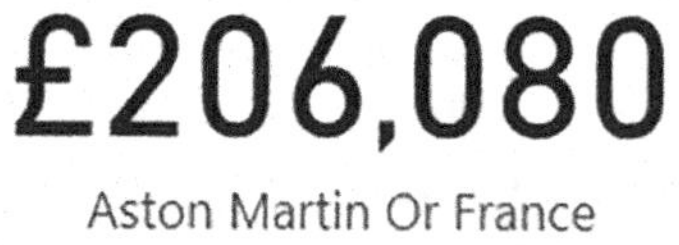

£206,080

Aston Martin Or France

Figure 16-21. *Total sale for an OR condition across two tables*

This output can, of course, be filtered or sliced perfectly normally in Power BI Desktop.

What this code does is

1. Use CROSSJOIN() to create the Multicriteria table variable. This table contains two columns - CountryName and Make. All possible combinations of the values in the two fields are in this table. You can see the first few records of this table in Figure 16-22.

CountryName	Make
Germany	Alfa Romeo
Belgium	Alfa Romeo
Switzerland	Alfa Romeo
United States	Alfa Romeo
France	Alfa Romeo
Spain	Alfa Romeo
Italy	Alfa Romeo
United Kingdom	Alfa Romeo
Germany	Aston Martin
Belgium	Aston Martin
Switzerland	Aston Martin
United States	Aston Martin
France	Aston Martin
Spain	Aston Martin
Italy	Aston Martin
United Kingdom	Aston Martin

Figure 16-22. *The output from a CROSSJOIN() function*

2. Create a table variable named SelectedCriteria that applies an OR filter to the Multicriteria table variable using the FILTER() function. As you can see from Figure 16-23, the output includes either Aston Martins or vehicles sold to clients in France.

CountryName	Make
France	Alfa Romeo
Germany	Aston Martin
Belgium	Aston Martin
Switzerland	Aston Martin
United States	Aston Martin
France	Aston Martin
Spain	Aston Martin
Italy	Aston Martin
United Kingdom	Aston Martin
France	Austin
France	Bentley

Figure 16-23. *The output from a CROSSJOIN() function*

3. Aggregate sales by country, make, and model in the OrOutput table variable - which is filtered using the *SelectedCriteria* table variable. Here, the table variable is used instead of applying a FILTER function inside the SUMMARIZECOLUMNS() function as was done in previous examples.

4. Finally the OrOutput table variable is processed by an aggregator function to return a scalar value.

CROSSJOIN(), then, creates what is called a "Cartesian" result where all possible combinations of the values from the two source columns are produced. In most cases this kind of output is an intermediate step in DAX that is used to deliver other outputs.

Table Functions

DAX provides a swathe of table functions, and it is simply not possible to explain them all without the subject growing to cover an entire book. Table 16-1 briefly describes some of the remaining table functions that you may find useful when writing DAX.

Figure 16-1. *The Core Table Functions*

DISTINCT()	Returns a set of unique elements - removing duplicate rows if applied to a table and duplicate values if applied to a column as well as ignoring blanks when getting a column's distinct values
GENERATE()	Returns all the rows from a second table applied to each row of the first
NATURALINNERJOIN()	Joins two tables using shared columns as the join fields
NATURALLEFTOUTER JOIN()	Joins two tables applying a SQL-like outer join using shared columns as the join fields
SUMMARIZE()	An older version of SUMMARIZECOLUMNS() – best to use SUMMARIZECOLUMNS() instead (except in some edge cases)
VALUES()	Returns a table (list) of distinct values

Conclusion

This chapter introduced you to DAX table functions. This small family of DAX functions has the capability to open up disproportionally wide horizons when it comes to creating extremely targeted analytics.

Firstly you met the SUMMARIZECOLUMNS() table function. This function lets you create ad-hoc virtual tables that can be grouped, filtered, and aggregated as well as all combinations of these three possibilities. You can then use the virtual table in further calculations either as a filter or to allow further calculations.

Then you discovered how to narrow or widen the output from DAX tables using the SELECTCOLUMNS() and ADDCOLUMNS() functions. You topped this off by using the CALCULATETABLE() function to add modifiers and filters to table functions.

The INTERSECT(), UNION(), and EXCEPT() functions. These three table functions allow you to detect – or exclude - the overlap between virtual table data as well as creating virtual tables from multiple table expressions.

As a final flourish you saw the CROSSJOIN() function. This function allows you to create a Cartesian product of all the specified columns. One use for this is to create table variables that allow OR logic to be applied across different tables.

All the DAX in this chapter can be found in the Power BI Desktop file PrestigeCarsDimensionalWithTables.pbix. This file can be downloaded along with all the other sample files from the Apress website.

It is now time to move on to the penultimate chapter in your introduction to DAX. In the next chapter, you will learn how to override and go beyond the limits of the data model using DAX.

CHAPTER 17

Beyond the Data Model

Up until now this book has taught you how to create DAX that is built upon (and uses) a basic Power BI data model. Essentially this meant

- Only using data that was imported as the basis for calculations
- Using the table relationships that were created in the initial data model
- Using the filter direction that was defined in the data model

While it is perfectly possible to solve most (and in some cases all) of your analytical challenges in Power BI using only a "classic" data model, there will inevitably be times when you will need to go beyond the limitations of the existing data model. This chapter is an introduction to some of the techniques that you can use to

- Add small amounts of data to a data model using DAX
- Refer to data in unconnected tables
- Override the filter direction between connected tables

As you progress with DAX and the Power BI data model, you could need to apply ever more complex solutions to analytical challenges. It is clearly impossible to provide all the answers, but the aim of this chapter is to provide an understanding of some of the ways that you can use DAX to extend and go beyond the limits of a data model.

This chapter uses the Power BI Desktop file PrestigeCarsDimensional2.pbix that you can find with the source files for this book.

419

© Adam Aspin 2023
A. Aspin, *Pro DAX and Data Modeling in Power BI*, https://doi.org/10.1007/978-1-4842-8995-2_17

Adding Data

Sometimes the source data that you are using does not contain all the information that you need. It may be that you require some high-level data to extend the data model. Maybe you want to store some values that you use regularly in DAX calculations. It could be that you want to add fixed data that you do not want to update. Whatever the reason, DAX allows you to specify ad-hoc data tables that you can add to the data model. There are essentially two ways to do this:

- The table constructor

- The DAX DATATABLE() function

Let's look at these in turn.

Note   These techniques are only appropriate for small amounts of data. They should never be used to replace a properly structured data ingestion process. The data can, however, be modified in Power BI Desktop fairly easily.

The DAX Table Constructor

In the companion volume "Pro Data Mashup for Power BI," you learned how to ingest data from multiple data sources as the basis for your analytics. On some occasions you may need to add custom data tables to the data model that are not linked to outside sources. Reasons for this could be (among many other potential drivers)

- You want to define a small sample table of data to test or prepare data structures.

- You need to ensure that data is never changed - by accident or design - even if the source data is modified.

- Your data source is not yet fully implemented but you need to have data available in order to continue data model development.

In cases like these you can create tables of data in DAX using the *table constructor*. Here is how you can do this.

1. Open the file PrestigeCarsDimensional2.pbix.

2. In the Home ribbon click New table.

3. In the Formula Bar enter the following code:

```
YearlyBudget =
    {
        (DATE(2019, 12, 31), CURRENCY(3500000)),
        (DATE(2020, 12, 31), CURRENCY(4500000)),
        (DATE(2021, 12, 31), CURRENCY(500000)),
        (DATE(2022, 12, 31), CURRENCY(7500000)) ,
        (DATE(2023, 12, 31), BLANK())
    }
```

4. Confirm the DAX by clicking the tick icon or pressing Enter.

5. Rename the columns (which will be called Value1, Value2, etc.). In this example I have named them "Date" and "Amount."

You can see both the DAX and the table that it creates in Figure 17-1.

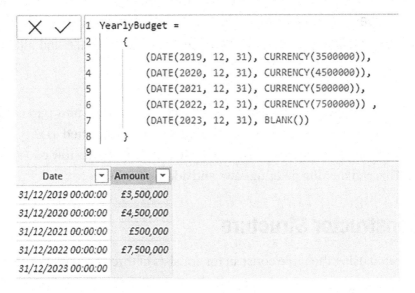

Figure 17-1. A table created using the table constructor

There are a few core points to note if you use the table constructor to create a "hard coded" data table:

- You have to rename the columns individually in the Fields list.

- You can use DAX functions to create the data elements. This can include calculating column data.

- All rows (records if you prefer) that are added using the table constructor *must* have the same number of elements. This means that you cannot leave any element (or column if you prefer) empty.

- To add an empty element, use the BLANK() keyword.

- Power BI will guess the data type from the data in each column. However you can override this by choosing an appropriate data type using the data type option of the Column tools ribbon.

Note   While it is possible to use the Enter data button in the Home ribbon to create tables of data, this approach (whether manually typing data or copying and pasting) is limited to 1500 cells of data. The row constructor has no such limitation. In practice, however, you wouldn't want to maintain this many lines of code – you should consider separating out that reference data into a csv file and ingesting that separately.

The data in any tables created using the table constructor are now part of the data model. They will never be updated when the data model is refreshed as they are not connected to any source. You can nonetheless alter the data in the row constructor DAX at any time. This can include deleting rows and adding further rows.

Table Constructor Structure

A table generated using the table constructor consists of three principal elements:

- The table name
- Outer curly braces
- Multiple rows of data

Inside each row of data, the following structure must be respected:

- Each row of data must be enclosed in parentheses.

- Separate each value (or column) using a comma.

- End each row with a comma.

Concerning the data elements, you need to apply the following guidelines:

- Text should be entered in double quotes.

- Numbers do not need quotes and can (inevitably) contain decimals but should *not* be formatted. That includes thousands separators and currency symbols. Numbers can be wrapped in functions such as VALUE().

- Dates should be entered in a recognized date or datetime format inside double quotes. Alternatively the DATE() or DATEVALUE() functions or the dt"2022-09-1" syntax can be used.

- Times should be entered in a recognized time format inside double quotes. Alternatively the TIME() or TIMEVALUE() functions can be used.

- Boolean values should be entered as TRUE() or FALSE().

- BLANK() must be used if an element would otherwise be empty.

DATATABLE()

Another (and more powerful approach) to creating hard-coded tables inside a data model is to use the DATATABLE() function. While the result is identical to creating a table using a row constructor, the DATATABLE() approach allows you to define the column names and data types as part of the table definition.

1. Switch to Table view and click the New Table button in the Table tools ribbon.

2. Enter the complete contents of the file File BudgetTable.txt that you can find in the folder containing the downloaded sample data.

3. Confirm the DAX by clicking the tick icon or pressing Enter.

Figure 17-2 shows part of the resulting table.

Year	Color	Amount	YearEnd
2019	Black	350000	31/12/2019 00:00:00
2019	Blue	250000	31/12/2019 00:00:00
2019	British Racing Green	125000	31/12/2019 00:00:00
2019	Canary Yellow	100000	31/12/2019 00:00:00
2019	Dark Purple	25000	31/12/2019 00:00:00
2019	Green	180000	31/12/2019 00:00:00
2019	Night Blue	75000	31/12/2019 00:00:00
2019	Pink	25000	31/12/2019 00:00:00
2019	Red	150000	31/12/2019 00:00:00
2019	Silver	90000	31/12/2019 00:00:00

Figure 17-2. *A table created with the DATATABLE() function*

Figure 17-3 shows the initial structure of the data as found in the sample file.

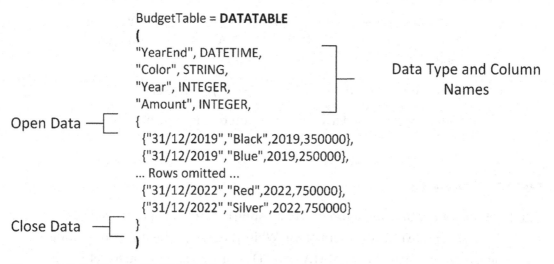

```
BudgetTable = DATATABLE
(
"YearEnd", DATETIME,
"Color", STRING,            Data Type and Column
"Year", INTEGER,                   Names
"Amount", INTEGER,
Open Data     {
    {"31/12/2019","Black",2019,350000},
    {"31/12/2019","Blue",2019,250000},
    ... Rows omitted ...
    {"31/12/2022","Red",2022,750000},
    {"31/12/2022","Silver",2022,750000}
Close Data    }
)
```

Figure 17-3. *The DATATABLE() function explained*

As a DAX function, DATATABLE() uses the following two parameters:

- The data specification (column name and data type). This is also called the data header.

- The data records.

The column names specified in the data header must be enclosed in double quotes. Inside each row of data the following structure must be respected:

- Each row of data must be enclosed in curly braces.

- Separate each column using a comma.

- End each row with a comma.

- Data must be simple, raw data. That is, no DAX expressions can be used.

- To add an empty element, use the BLANK() keyword.

- DATATABLE() will add BLANK() automatically (and invisibly) if there are missing values for a column.

- The available data types are

  - STRING

  - INTEGER

  - DATETIME

  - CURRENCY

  - DOUBLE

  - BOOLEAN

Many to Many Relationships in the Data Model

Ad-hoc tables created using DATATABLE() or a row constructor can be integrated into an existing data model just as imported tables can. After all, tables created in DAX are simply data tables.

However, I want to take the budget table as an example of how a data model can be made to handle tables of data where the relationship between tables is not as simple as the simple one to many joins that are used in the Prestige cars dimensional data model.

If you look at the data in the BudgetTable table, it has

- A year field that should somehow be connected to the Date dimension

- Color data that needs to connect to the Vehicle dimension (where color data is stored)

The challenge with the BudgetTable table is that it contains data at an aggregate level. This means that the same data appears in the Year field multiple times, as do elements in the Color field. As a consequence of this, it is *not* possible to have unique key fields that allow the kind of simple joins that underpin the other relationships between tables in the rest of the data model. So a little creativity is required to include budget data.

The first trick is inside the BudgetTable table itself. Every year has a year end date, and this corresponds to a unique date in the Date table. So it is easy to add 31st December to each year in the BudgetTable table to produce a new column with the year end date. This can be done wherever the data is produced (often in Excel in my experience of dealing with budget data).

The second trick concerns the color data. Fortunately, Power BI allows you to join tables where the same data appears many times in each table. Remember that the same color values will appear many times in the Vehicle table for each combination of color, make and model, etc.

Let's see this in action.

1. Switch to Model view.

2. Drag the Vehicle, BudgetTable, and Date tables to organize them (purely as a visual aid) as shown in Figure 17-5.

3. Drag the YearEnd field from the BudgetTable table on to the DateKey field in the Date table. A "classic" one to many join will be created.

4. Drag the Color field from the BudgetTable table over the Color field in the Vehicle table. The Create Relationship dialog will appear.

5. Select Single (Vehicle filters Budget table) as the cross filter direction. The Create Relationship dialog will look like Figure 17-4.

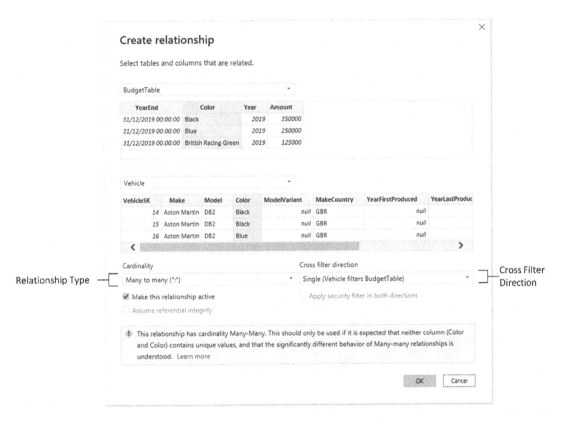

Figure 17-4. *The Create Relationship dialog for a many to many join*

The Model view (for these three tables) will look something like Figure 17-5.

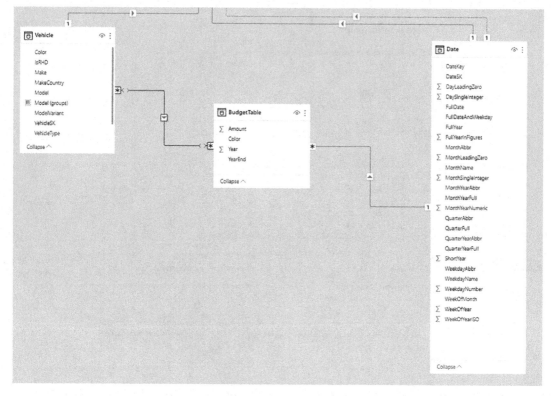

Figure 17-5. *Model view for a many to many join*

You can see in Figure 17-4 that a many to many relationship like the one between the Vehicle and BudgetTable tables has

- Asterisks at each end of the join line to indicate that this is a many to many join

- A single triangle representing the filter direction that you selected at step 5

- Double chevrons at each end of the join line

You can now use the budget data in Power BI visuals completely normally. An example of this is the table in Figure 17-6.

FullYear	Color	SalePrice	Amount
2019	Black	£374,490.00	350000
2020	Black	£1,251,410.00	1250000
2021	Black	£3,087,550.00	2500000
2022	Black	£1,983,695.00	2000000
2019	Blue	£284,600.00	250000
2020	Blue	£423,650.00	400000
2021	Blue	£688,920.00	500000
2022	Blue	£1,186,400.00	1500000
2019	British Racing Green	£111,500.00	125000
2020	British Racing Green	£663,250.00	600000
2021	British Racing Green	£872,340.00	750000
2022	British Racing Green	£566,385.00	500000
2019	Canary Yellow	£132,000.00	100000
2020	Canary Yellow	£287,070.00	250000
2021	Canary Yellow	£79,750.00	50000
2022	Canary Yellow	£134,045.00	100000
2019	Dark Purple		25000
2020	Dark Purple	£283,000.00	250000
2021	Dark Purple	£219,175.00	100000
Total		**£21,977,950.00**	**20420000**

Figure 17-6. Using data from multiple levels of granularity in the source tables

---

Note   When combining data at different levels of granularity, you will probably find yourself displaying the data only at the higher level of granularity.

Avoiding Many to Many Relationships

There may be cases where you do not want to use many to many relationships in your data model. The core reasons for this could be that

- The output will not include a blank row corresponding to mismatched rows in the other table

- The output will not handle rows where the column used in the relationship in the other table is blank

- RELATED() cannot be used in DAX formulas across many to many joins, as more than one row could be related

- Using ALL() in DAX formulas does not remove filters that are applied to other, related tables by a many-to-many relationship

Whatever the reasons for wanting to avoid a many to many join, there is one standard workaround. This is to add a further simple table that contains one record for each of the values that appear multiple times in the fields that would be used in a many to many join. Here is how you can do this:

1. Switch to Table view.

2. In the Table tools menu, click New table and add the following DAX snippet: this will create a table containing one row for each available color.

    ```
    ColorFilter = VALUES('Vehicle'[Color])
    ```

3. Switch to Model view.

4. Delete the join between the BudgetTable and Vehicle tables if you added it as described in the previous section.

5. Place the ColorFilter table below the BudgetTable and Vehicle tables (this is not strictly necessary but it helps when creating the joins).

6. Drag the Color field from the ColorFilter table on to the Color field in the BudgetTable table.

7. Drag the Color field from the ColorFilter table on to the Color field in the Vehicle table.

You can see the new data model in Figure 17-7.

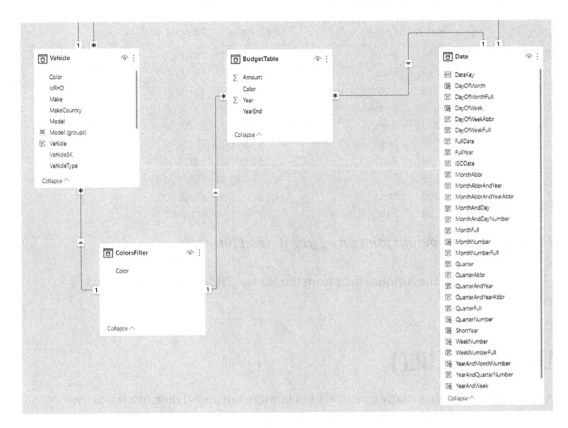

Figure 17-7. *An aggregate level filter table*

Note   The VALUES() function returns a unique list of the elements in a column.

At this point the data model contains three Color fields (one in the Vehicle table and one in the Budget table and one in the ColorFilter table) that can be used for displaying and filtering data. When filtering the Sales table (the main fact table), you can use either of the color fields. However if you are displaying or filtering data from the BudgetTable table, it is *vital* to use the color field from the ColorFilter table. This is because only this field is joined to the BudgetTable table as well as being joined to the Sales table (via the Vehicle table).

To make this clearer, Figure 17-8 shows what happens if you use the Color field from the Vehicle table to compare and contrast values from the Sales and BudgetTable tables.

FullYear	Color	SalePrice	Amount
2019	Black	£374,490.00	1370000
2019	Blue	£284,600.00	1370000
2019	British Racing Green	£111,500.00	1370000
2019	Canary Yellow	£132,000.00	1370000
2019	Green	£216,600.00	1370000
2019	Night Blue	£57,990.00	1370000
2019	Red	£154,795.00	1370000
2019	Silver	£89,000.00	1370000
2020	Black	£1,251,410.00	4800000
2020	Blue	£423,650.00	4800000

Figure 17-8. The output from an aggregate level filter table

As you can see, the Amount field from the BudgetTable table is not filtered in this case.

LOOKUPVALUE()

A key function that has many practical uses in more advanced data models is the LOOKUPVALUE() function. This value can be used to

- Avoid hardcoded values in DAX measures and calculated columns

- Replace table joins in a data model in some circumstances

Let's look at these in turn.

Avoid Hardcoded Values

In previous chapters you learned the basics of DAX arithmetic. Calculations in DAX frequently involve hard-coded values (sales taxes could be one example). However it is preferable in DAX - as in other programming languages - to avoid hard coding values wherever possible and refer to values stored in a reference table. This approach allows you to alter a core value once only and cascade the change through multiple formulas as and when updates to a reference value are required.

LOOKUPVALUE() is a simple and efficient solution to this kind of requirement. As an example, suppose that you have values for the following three elements that you wish to apply in multiple DAX calculations:

- Cleaning and Valeting Cost

- Oil Change Cost

- Export Costs

The first thing to do is to use the table constructor to create a simple DAX table named Fixed Costs table using the kind of DAX that you can see in the following short snippet:

```
FixedCosts =
    {
        ("CleaningAndValeting", 50),
        ("OilChange", 150),
        ("ExportCosts", 0.125)
    }
```

The first column has been renamed "CostElement," and the second "Cost."

Note   A table of fixed values could, of course, be stored outside the data model and loaded along with other data. However in this case let's use DAX table creation techniques for the reference data.

Now that you have a table of reference values, you can use LOOKUPVALUE() inside a DAX measure to refer to a selected element and use it in a calculation. The calculated column Margin Inc. Extras shows how this is done:

```
Margin Inc. Extras =
[SalePrice] - [CostPrice] + [SpareParts] + [LaborCost]
  - LOOKUPVALUE(
'FixedCosts'[Cost],
'FixedCosts'[CostElement],
"CleaningAndValeting"
)
```

This measure uses the data elements sale price, cost price, spare parts, and labor cost in a simple piece of arithmetic. The LOOKUPVALUE() function then adds the amount for cleaning and valeting.

LOOKUPVALUE() takes, at a minimum, three parameters:

- *Firstly*, the column (and table) that contains the reference value that you want to apply. This is the second column, named Cost in the table FixedCosts. This is the first column alphabetically in the fields list on the right hand side.

- *Secondly*, the column (and table) that contains the reference values that allow DAX to select the correct record in the lookup table. This is the first column, named CostElement in the table FixedCosts.

- *Thirdly*, the element that will be located in the column of reference values to identify the value to return. In this example this is "CleaningAndValeting."

So what you have done is told DAX "Go to the table FixedCosts and return the value found in the column Cost where the column CostElement is equal to CleaningAndValeting."

This formula is now dynamic and has avoided hard-coding the value for cleaning and valeting in the Margin Inc. Extras measure. Altering the reference value in the Fixed Costs table will update all formulas that use the lookup function including the formula that you just created.

You can see a more visual explanation of how LOOKUPVALUE() works in Figure 17-9.

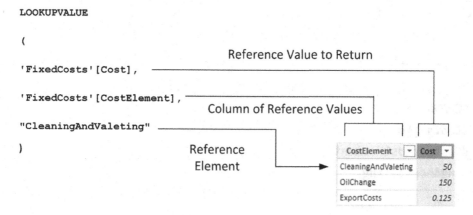

Figure 17-9. *The anatomy of the LOOKUPVALUE() function*

If you prefer, you can create measures that return a specific element using LOOKUPVALUE(). This avoids you having to copy and paste LOOKUPVALUE() formulas (or worse, rewrite them) several times. The measure that you create can then be used in other DAX formulas.

The following measure – SalesTaxPercentage - is an example of this approach.

```
ValetingPercentage =
LOOKUPVALUE('FixedCosts'[Cost], 'FixedCosts'[CostElement],
"CleaningAndValeting")
```

LOOKUPVALUE() is used in exactly the same way as before; only this time the measure simply returns the required value. So you can use this new measure, SalesTax, inside other measures as shown as follows:

```
CostIncValeting = SUM('Sales'[SalePrice]) * [ValetingPercentage]
```

You can see the output from the SalesTax measure in Figure 17-10.

FullYear	SalePrice	SalesTax
2019	£1,420,975.00	213,146.25
2020	£4,905,285.00	735,792.75
2021	£7,868,075.00	1,180,211.25
2022	£7,783,615.00	1,167,542.25
Total	£21,977,950.00	3,296,692.50

Figure 17-10. Using LOOKUPVALUE() in a measure to return a value

Imitate a Table Relationship

Another use for the LOOKUPVALUE() function is to imitate a table relationship. This can be practical in certain cases where you do not want to create a standard relationship between tables in a data model.

As an example of this approach, suppose that you have a budget table containing two columns, Year and BudgetValue, that is created using the following DAX:

```
YearlyBudget =
    {
        ("2019", CURRENCY(1500000)),
        ("2020", CURRENCY(4500000)),
```

```
        ("2021", CURRENCY(5000000)),
        ("2022", CURRENCY(7500000)),
        ("2023", BLANK())
    }
```

You can now add the budget for the year as a new column to the date table using LOOKUPVALUE() like this:

```
YearlyBudget =
LOOKUPVALUE(YearlyBudget[BudgetValue], YearlyBudget[Year],
'Date'[FullYear])
```

You can see the new column in Figure 17-11. In this case we have avoided creating a table join in the data model and using RELATED() to find the budget value.

MonthAbbrAndYear	YearAndQuarterNumber	YearAndWeek	YearAndMonthNumber	MonthAndDay	MonthAndDayNumber	YearlyBudget
Jan 2019	20191	201901	201901 Jan 5	0105	£1,500,000
Jan 2019	20191	201902	201901 Jan 6	0106	£1,500,000
Jan 2019	20191	201902	201901 Jan 7	0107	£1,500,000
Jan 2019	20191	201902	201901 Jan 8	0108	£1,500,000
Jan 2019	20191	201902	201901 Jan 9	0109	£1,500,000
Jan 2019	20191	201902	201901 Jan 10	0110	£1,500,000
Jan 2019	20191	201902	201901 Jan 11	0111	£1,500,000
Jan 2019	20191	201902	201901 Jan 12	0112	£1,500,000
Jan 2019	20191	201903	201901 Jan 13	0113	£1,500,000

Figure 17-11. Using LOOKUPVALUE() to replace a table join and RELATED()

In Figure 17-12 you can see how this data could be added to a table.

FullYear	SalePrice	YearlyBudget
2019	£1,420,975.00	£1,500,000
2020	£4,905,285.00	£4,500,000
2021	£7,868,075.00	£5,000,000
2022	£7,783,615.00	£7,500,000

Figure 17-12. Displaying the budget value using LOOKUPVALUE()

Lookup an Element Between a Range of Values

A classic Power BI data model joins one value in one table to the same value in another table. Yet there may be occasions where you have a value that you want to look up between a range of values in another table rather than having a direct correlation between identical values in both tables. This approach can also be used instead of hard-coding values inside a DAX column.

As an example of this, let's imagine that Prestige Cars needs to apply a scale of delivery charges for vehicle deliveries. The delivery cost varies according to the sale price of the car. A simple custom DAX table contains the scale of charges. This table is defined using the following DAX:

```
DeliveryCharges =
DATATABLE(
 "StartRange", INTEGER,
 "EndRange", INTEGER,
 "Cost", INTEGER,
{
{0,50000,500},
{50000,100000,1000},
{100000,150000,2000},
{150000,10000000,5000}
}
)
```

Figure 17-13 shows the table that this DAX creates.

StartRange	EndRange	Cost
0	50000	500
50000	100000	1000
100000	150000	2000
150000	10000000	5000

Figure 17-13. *A table for range lookups*

Now that we have a lookup table containing ranges of vehicle sales prices - and the associated delivery cost for each set of range thresholds - you can create a measure to return the appropriate delivery cost for each individual car sale. The following DAX is one way to do this.

```
DeliveryCosts =
VAR _SalePrice = SUM('Sales'[SalePrice])
RETURN
MAXX(
    FILTER('DeliveryCharges',
    _SalePrice >= 'DeliveryCharges'[StartRange]
    &&
    _SalePrice <= 'DeliveryCharges'[EndRange]
    ),
    [Cost]
)
```

Let's see exactly how this measure works. This measure is intended to be used when a list of vehicles is output.

- Firstly, a variable is used to find the sale price of a vehicle. This variable finds the sale price of each car in a list of sales.

- The FILTER() function then takes the _SalePrice variable and compares it to the lower and upper threshold values in the DeliveryCharges table using a simple pair of comparison functions.

- The MAXX() function iterates over the virtual table created by the FILTER() function and returns the contents of the Cost column that correspond to the row that has been isolated by the FILTER() function.

One application of a range lookup would be to add the DeliveryCosts measure as a new column to the Sales table. This is as simple as adding the following DAX as a new column.

```
Delivery = [DeliveryCosts]
```

Figure 17-14 shows the first few rows for a table that lists vehicle sales. You can see that the appropriate delivery cost is returned for each sale price.

FullDate	StockCode	SalePrice	Delivery
02 January 2019	2189D556-D1C4-4BC1-B0C8-4053319E8E9D	£19,950.00	500
02 January 2019	B1C3B95E-3005-4840-8CE3-A7BC5F9CFB3F	£65,000.00	1000
25 January 2019	A2C3B95E-3005-4840-8CE3-A7BC5F9CFB5F	£220,000.00	5000
03 February 2019	558620F5-B9E8-4FFF-8F73-A83FA9559C41	£19,500.00	500
16 February 2019	72443561-FAC4-4C25-B8FF-0C47361DDE2D	£11,500.00	500
14 March 2019	C1459308-7EA5-4A2D-82BC-38079BB4049B	£29,500.00	500
24 March 2019	E6E6270A-60B0-4817-AA57-17F26B2B8DAF	£49,500.00	500
30 March 2019	CEDFB8D2-BD98-4A08-BC46-406D23940527	£76,000.00	1000
04 April 2019	D63C8CC9-DB19-4B9C-9C8E-6C6370812041	£36,500.00	500
06 April 2019	6081DBE7-9AD6-4C64-A676-61D919E64979	£19,600.00	500
30 April 2019	4C57F13A-E21B-4AAC-9E9D-A219D4C691C6	£80,500.00	1000

Figure 17-14. *Returning a value from between a range of values in a lookup table*

USERELATIONSHIP( )

As I have mentioned (and as you have no doubt discovered when building data models in Power BI Desktop), it is impossible to have more than one active relationship between two tables at once. You can create more than one relationship between two tables; that is not a problem as Power BI will let you do this if the relationship is valid. However, only one of the relationships will be used.

This can be a limitation in certain circumstances. A classic scenario is when you want to use a date table to display date elements from multiple date fields in a single fact table. To make this clearer and more intuitive, take a look at the file PrestigeCarsDimensional2.pbix. Here we are going back to a very simple data model to avoid unnecessary complication. You can see that this data model contains the DateDelivered Column in the Sales table as well as the DateSK field that is used to enable time intelligence - and that has always been used as the basis for the join to the date table to enable time analysis and time intelligence relative to the date of sale.

Yet it is perfectly plausible that you may need to display date information using the date table (or apply time intelligence) when analyzing by the *delivery date* rather than the sale date.

Instead of modifying the active relationship in the data model itself (which would only reverse the problem), switching the analysis to a different data field is perfectly feasible if you learn to apply the DAX USERELATIONSHIP() function. This is a two-step process.

- Firstly you need to join the second date field to the date table. This is as simple as dragging one field from one table over the corresponding join field in the other table.

- Secondly you write a DAX measure to force the calculation to use the new relationship.

In this data model creating a second, inactive relationship between the DateKey field in the Date table and the DateDelivered field in the Sales table will give you a join that looks like the one shown in Figure 17-15.

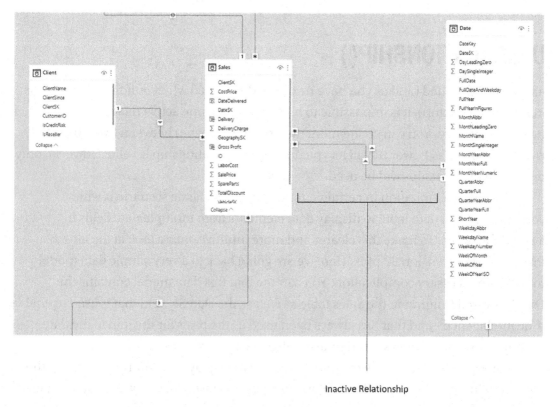

Inactive Relationship

Figure 17-15. *An inactive table join*

This relationship is shown as a dotted line. The initial join remains active and functions normally. The new relationship is present but not active and will not be used unless specifically applied.

Now that an inactive relationship is set up, it can be used when required. This entails creating DAX measures that apply the USERELATIONSHIP() function as a CALCULATE() modifier.

As an example, suppose that you want to calculate the sales by delivery date. This means writing the following measure:

```
TotalSalesAtDeliveryDate =
CALCULATE(SUM('Sales'[SalePrice]), USERELATIONSHIP('Date'[DateKey],
'Sales'[DateDelivered]))
```

If you compare this measure with the (extremely simple) measure that you created many chapters ago to calculate total sales, you can see that there is only a little added complexity. A measure that uses the active relationship is

```
TotalSales = SUM('Sales'[SalePrice])
```

This core measure is extended to use a different join by

- Wrapping the same aggregation inside CALCULATE()

- Adding USERELATIONSHIP() as a modifier

- Specifying inside the USERELATIONSHIP() function the two fields that make up the currently inactive relationship. These fields can be in any order.

You can compare the output from the initial and new sales calculation measures in Figure 17-16.

FullYear	MonthAbbr	TotalSales	TotalSalesAtDeliveryDate
2019	Jan	£304,950	£304,950
2019	Feb	£31,000	£31,000
2019	Mar	£155,000	£79,000
2019	Apr	£145,100	£132,100
2019	May	£373,450	£462,450
2019	Jun	£22,950	£22,950
2019	Jul	£107,185	£107,185
2019	Aug	£5,500	£5,500
2019	Sep	£37,200	£37,200
2019	Oct	£146,190	£22,600
2019	Nov	£22,950	£146,540
2019	Dec	£69,500	£69,500
2020	Jan	£451,150	£451,150
2020	Feb	£171,040	£167,390
2020	Mar	£220,500	£224,150
2020	Apr	£258,750	£102,950
2020	May	£99,030	£155,800
2020	Jun	£237,300	£336,330
2020	Jul	£608,240	£528,740
Total		**£21,977,950**	**£21,977,950**

Figure 17-16. Forcing a measure to apply an inactive relationship with USERELATIONSHIP()

Note It is not just date tables that can be used to apply inactive relationships. Indeed any inactive relationship between two tables can be used in this way.

CROSSFILTER ()

In Chapter 5 you saw that setting up cross filtering correctly (i.e., defining which table can filter other tables) is one of the core elements of a good data model. Cross filter direction is a fundamental aspect of the data model and often shapes the structure of the model.

Fortunately DAX allows you to add some flexibility to this aspect of the data model. More specifically, you can add measures that override the cross-filter direction of an existing relationship inside the measure that you create. This is nearly always done through applying the CROSSFILTER() function as a modifier to the CALCULATE() function.

This is best appreciated with an example. If you take a look at Figure 17-17, you can see part of the Prestige Cars data model with two dimensions linked to a fact table.

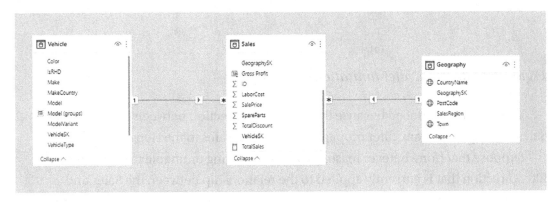

Figure 17-17. *Part of the Prestige Cars data model*

In this scenario one way filtering has been applied. This means that the Vehicle and Geography tables can filter the Sales table, but neither of these dimensions can filter the other dimension. To confirm this, create the following simple measure and add it to the Geography table. This measure tries to display the number of countries where each color is sold.

```
Number Of Countries = DISTINCTCOUNT('Geography'[CountryName])
```

Creating a table of colors that attempts to show how many countries have sales for each color gives the output shown in Figure 17-18.

Color	Number Of Countries
Black	8
Blue	8
British Racing Green	8
Canary Yellow	8
Dark Purple	8
Green	8
Night Blue	8
Pink	8
Red	8
Silver	8
Total	**8**

Figure 17-18. *Cross filter limitations*

This result is produced because the cross filter direction between the Geography and Sales tables prevents any filter propagation from the Sales to the Geography table.

Suppose that (for whatever reason) you are unwilling or unable to change the filter direction that is currently applied to the relationship between the Sales and Geography table. Yet you need to be able to output the number of countries where each color is sold. This can be solved by creating the following measure (also added to the Geography table).

```
Correct Number Of Countries =
CALCULATE(
    DISTINCTCOUNT('Geography'[CountryName]),
    CROSSFILTER('Sales'[GeographySK],
                'Geography' [GeographySK],Both)
)
```

The output from this measure is shown in Figure 17-19.

Color	Correct Number Of Countries
Black	8
Blue	6
British Racing Green	7
Canary Yellow	4
Dark Purple	6
Green	8
Night Blue	5
Pink	5
Red	7
Silver	8
Total	**8**

Figure 17-19. *Using CROSSFILTER() as a CALCULATE() modifier*

As you can see from the previous DAX snippet, CROSSFILTER() has been added as a CALCULATE() modifier to the original measure. CROSSFILTER() itself required three parameters:

- *Firstly,* one of the tables in an existing relationship
- *Secondly,* the other table in an existing relationship
- *Finally,* the cross filter direction to be applied inside this measure

The five types of possible cross filter override are given in Table 17-1.

Table 17-1. *CROSSFILTER() direction options*

Crossfilter Type	Description
None	No cross filtering is allowed
Both	Either side can filter the other
OneWay	Filters on the "one" side can apply to the table on the "Many" side
OneWay_ LeftFiltersRight	The table that is set as the first parameter can filter the table set as the second parameter
OneWay_ RightFiltersLeft	The table that is set as the second parameter can filter the table set as the first parameter

There are a few points to note when using CROSSFILTER():

- CROSSFILTER() applies only inside the measure where it is used.

- If you are traversing multiple joins to filter tables that are not directly connected, you can add multiple CROSSJOIN() modifiers inside the CALCULATE() - or other appropriate - function to specify the crossjoin type for each relevant join. Each CROSSJOIN() modifier is separated by a comma from any other CALCULATE() modifiers.

- CROSSFILTER will override any existing cross-filtering setting in a relationship.

- CROSSFILTER() can only be applied to functions that take a filter as an argument. This means CALCULATE, CALCULATETABLE, CLOSINGBALANCEMONTH, CLOSINGBALANCEQUARTER, CLOSINGBALANCEYEAR, OPENINGBALANCEMONTH, OPENINGBALANCEQUARTER, OPENINGBALANCEYEAR, TOTALMTD, TOTALQTD, and TOTALYTD functions.

Conclusion

This chapter introduced you to some of the ways that you can use DAX to extend the data model. First, you saw how to add small amounts of data without needing to import source data. Then you learned how to reference data across tables without table joins using both direct and range lookup techniques.

You then saw how to implement multiple joins between tables and use DAX to apply inactive joins to deliver results. Finally you discovered how to override the filter direction in table joins.

These techniques are only the basics of all that DAX can do to extend and override the core data model. A lot more is possible, of course. However you have now learned the basics of applying DAX to make your data models more supple and better able to deliver more of the analytics that you need.

The techniques used in this chapter can be found in the sample file PrestigeCarsDimensionalWithDataModelExtensions.pbix.

It is now time to move on to the final chapter in this book. This is where you will learn about filter context in DAX expressions.

CHAPTER 18

Evaluation Context

When developing DAX formulas in this book, you have essentially seen how to produce formulas that deliver the results that you expect. Now that you have seen the *how* – and to conclude your introduction to DAX - it is time to move on to an initial understanding of *why* DAX formulas work as they do.

This is because, when working with DAX (at least if you want to develop any complex formulas), you need to understand *Evaluation Context*. This is the basis of the dynamic data analysis that you are learning to deliver using Power BI Desktop. It is the basis of the approach where the results of a formula can change to reflect the current row or cell selection and also any related filters.

You have been using evaluation context in all the DAX formulas in this book as well as in any DAX formulas that you have ever written. Sometimes, however, the focus (especially when starting out with DAX) is simply to make something work. This is perfectly normal and understandable. However, when faced with more complex DAX challenges, it is often easier to deliver a functioning formula if it is based on a clear understanding of the way that DAX works. As evaluation context is quite probably the single most fundamental concept in DAX, I think that it is worth looking back at the DAX that you have learned in this book and explaining how all of it - every single formula - is underpinned by the concept of evaluation context. This way, once you have a clearer grasp of this core concept, you will find it easier to deliver more complex formulas.

Evaluation context can become an extremely complicated subject. However, as this is an introductory book on DAX, I am deliberately simplifying some of the ideas explained below. After all, the aim is to get you started with DAX, and not to scare you off right at the start.

The two key elements that you need to understand are

- Row context
- Filter context

Let's take a brief look at these in turn.

© Adam Aspin 2023
A. Aspin, *Pro DAX and Data Modeling in Power BI*, https://doi.org/10.1007/978-1-4842-8995-2_18

Row Context

Row context is essentially the values from the current row. You saw this when creating new columns in a table. This means that any fields that you used in a calculation always used other fields from the same record - or virtually added to the current row from a linked table.

The great thing about Row context is that it is largely automatic and will be handled for the most part by DAX without any intervention on your part (unless you choose to override it - of which more later). This is because row context is another way of saying "iterate over the entire table and carry out a calculation for the current row only." In other words, row context can be taken to mean "current row only." Indeed, row context is why you can write simple formulas such as the first DAX formula that you saw in this book. It was a new column added to the Vehicle table that looked like this:

```
Vehicle = [Make] & " " & [Model]
```

As you saw when creating this formula, you did not need to specify the table that contains the fields Make and Model. This is because the concept of row context is applied to limit the formula to the current table (or calculations that use RELATED() or RELATEDTABLE() to look beyond the current table while always referring to the current row). A corollary of this limitation to the current table is that the formula is applied (or iterates over) each row in the table.

Row context does not limit calculations to only the current table. However - and as you saw in Chapter 5 - you need to force the calculation to traverse the data model if you wish to use data from other tables. This is why you needed to use the RELATED() and RELATEDTABLE() functions for more complex calculations.

So the core takeaways concerning row context are

- Row context applies to the current table

- Row context is limited to the current table (whether it be Real or virtual) unless you specify a link to other tables using use the RELATED() and RELATEDTABLE() functions (and that the data model has a join path between the core table and the related table)

- Row context iterates over the entire table-ensuring that a formula is applied to the entire column(s)

- Row context only sees the current row when it iterates over a table

- Row context cannot see the entire data model

- Row context is unaware of filter context (which is explained next)

Row Context Beyond Tables

However, row context is not limited to columns that you add to DAX tables. It also applies to

- Iterator functions (like SUMX())

- FILTER()

- Virtual tables

When using these DAX elements, the principles of row context also apply. So, when iterating over a virtual table using a function such as AVERAGEX(), you need to be aware that - as is the case with standard tables - the principles of row context that are outlined in the previous section also apply.

Filter Context

Filter Context is a more powerful and much more vast concept than row context. Put succinctly, filter context

- Is aware of the entire data model

- Is not limited to a single table or row

- Underpins how measures work

- Can be overridden using the CALCULATE() and CALCULATETABLE() functions

Initially - and out of the box - filter Context is the combination of the following factors that produces a calculated result:

- Report-level filters

- Page-level filters

- Visualization-level filters

- Slicers

- Interactive selection

- Row and Column context

The first three are cumulative (report-level filters, page-level filters, and visualization-level filters) and reduce the available data that a visualization can show. For the moment just consider them as a set of filters that only allows certain data to be used. Slicers and interactive selection then narrow down the output data by extending the range of filters that are applied. Row and column context are best thought of as the row and column headers in a pivot table (or a crosstab if you prefer) or, similarly, the "x" axis and legend in a bar chat/histogram, for example. These define the intersections of data that can be shown.

These concepts were first introduced in Chapter 8 and explained in Figure 8-11. I will not reproduce this figure here, but suggest that you refer to it if you need a visual explanation of the cumulative effect of filter context.

However, what I have just described is the *implicit* filter context. This is the initial filter context that is generated automatically by Power BI. This will be applied to any measure that you use in a visual. You are probably used to implicit filter context as (at least when starting out with Power BI) you relied on it to deliver the correct aggregations in your tables, matrices, and charts. This is what makes standard visuals "just work."

Overriding the Initial Evaluation Context

The implicit filter context works beautifully when you want to deliver classic aggregations that apply to all the data specific column (or columns) and where the filtering can be delivered using filters, slicers, rows, and columns (or axes in charts).

DAX gets more interesting (and yes, this is a polite way of saying that it is also more challenging) when the implicit filter context no longer gives you the results that you want. This is where you need to override the implicit filter context to push DAX to the next level. Chapters 9 through 17 (as well as one section in Chapter 5) introduced you to some of the main functions that you can use to take the initial, automatically generated filter context and adjust it to give the specific results that you require.

Overriding the Row Context

If you cast your mind back to Chapter 5 (or if you flip back and review it), you will remember that we created the following formula as a new column in the Geography table

```
Number of Sales = Count(Sales[ClientSK])
```

This formula was perfect - if it was the total number of sales that was required and that this same figure needed to be applied in each row of the table. This is because *row context is utterly unaware of filter context*. So this figure will remain the same in all output rows in visuals no matter what filters are applied.

However, row context can be overridden. You saw this also (albeit briefly) in Chapter 5 in the following formula:

```
Number of Sales Per Postcode = CALCULATE(COUNT(Sales[ClientSK]))
```

This formula overrode the row context and switched to a filter context. This meant that the formula could see the entire data model and use table relationships. Specifically, the relationship between the Geography and Sales tables allowed the implicit filter between the two tables to be applied. The output was that it is no longer the grand total of the number of sales that is displayed in a new column, but the sales relative to that row. This effect is called *context transition*.

Overriding the Implicit Filter Context

Just as CALCULATE() and CALCULATETABLE() override row context by switching to "full" filter context, they allow you to override any initial filter context and apply exactly the filtering that you require in a measure.

As I do not want to regurgitate all that you saw in Chapters 10 through 17, let's resume the effects of what filter context means when creating measures.

- Each value in a dashboard (whether this is a cell in a table or matrix or a data point in a chart) is a separate calculation that is evaluated completely separately from all other values.

- Each value starts as the result of an implicit filter context.

- The implicit filter context can then be overridden using the CALCULATE() and CALCULATETABLE() functions.

- CALCULATE() and CALCULATETABLE() allow you to add filters and overriders to go beyond the initial filter context.

In essence, the implicit filter context is the application of the data model to calculate each value. This is how Power BI is able to apply filters and slicers (depending on the filter direction set between tables) and output the result of these filter elements as well as row and column (or axis) filters.

Overriding the filter context consists of breaking the dependency on the implicit filter context. This is done by any or all of

- Temporarily re-defining relationships (as defined in Chapter 17)

- Setting a different filter direction for a calculation (also explained in Chapter 17)

- Applying CALCULATE() modifiers to break the limitations of implicit filters (outlined in Chapter 10)

How exactly you override and extend (or limit) the filter context is the basis of writing DAX formulas - as you have seen in many of the recent chapters. The principles are that DAX will

- Take the current filter context (the current "out of the box" interaction between filters, slicers, rows, columns, and axes)

- Remove filter elements that you specify (as explained in Chapter 10)

- Add filters that you specify (using filters inside the CALCULATE() function or the FILTER() function-as outlined in Chapter 11)

- Overlay the new filter context that you have defined on the initial filter context

Conclusion

This final chapter has given you the theory that underlies the practice of DAX calculations that you saw in Chapters 4 through 17. It does not attempt to explain all the multiple intricacies of filter context, but to give you a high level understanding of why DAX works the way it does.

This brings you to the end of not only the current chapter but also the book. I sincerely hope that you found this book useful and that you can use any knowledge gained to deliver efficient and elegant data models that include powerful DAX formulas to drive insight and deliver powerful dashboards.

APPENDIX

Sample Data

Sample Data

If you wish to follow the examples used in this book — and I hope you will — you will need some sample data to work with. All the files referenced in this book are available for download and can easily be installed on your local PC. This appendix explains where to obtain the sample files, how to install them, and what they are used for.

Downloading the Sample Data

The sample files used in this book are currently available on the Apress site. You can access them as follows:

1. In your web browser, navigate to the following URL: `github.com/apress/pro-powerbi-DAX-and-data-modeling`.

2. Click the button Download Source Code. This will take you to the GitHub page for the source code for this book.

3. Click Clone or Download ➤ Download Zip and download the file PowerBIDesktopSamples.zip.

You will then need to extract the files and directories from the zip file. How you do this will depend on which software you are using to handle zipped files. If you are not using any third-party software, then one way to do this is

1. Create a directory named `C:\PowerBIDesktopSamples`.

2. In the Windows Explorer navigation pane, click the file PowerBIDesktopSamples.zip.

3. Select all the files and folders that it contains.

4. Copy them to the folder that you created in step 1.

© Adam Aspin 2023
A. Aspin, *Pro DAX and Data Modeling in Power BI*, https://doi.org/10.1007/978-1-4842-8995-2

Index

A, B

Business intelligence (BI), 6, 87, 105, 129, 192, 327, 328

C

Calculated columns, 87
    approaches, 89, 91
    CALCULATE() function, 132
    cascading columns
        features, 111
        FORMAT() function, 113
        format option, 112–115
        graphical model, 111
        numeric format, 113, 115
        predefined currency formats, 114
    complex data model, 122, 123
    concatenation, 92–95
    COUNTROWS() function, 119
    cross filter direction
        data view, 126
        edit relationship dialog, 123, 124
        filter blocked, 125
        model view, 124
        modification, 124
        tables, 127
    dashboards, 90
    dashboard visualizations, 89
    data analysis, 117
    DAX functions, 88
    DAX variables, 376–378
    field relationship table, 118, 119

formula bar, 118
formula language, 87
formula selection, 93
functions, 127
geography table, 131
handling mistakes, 101
    modification, 102
    warning icon, 102
iterator functions, 276, 280, 281
learning process, 117
limitations, 132, 133
Lookup() function, 119
math operators, 105–108
measures (*see* Measures)
options, 116
PowerPivot, 104
reference elements, 119, 120
RELATED() function, 121
RELATEDTABLE() function, 119
rename option, 95
round and truncate values, 107–110
statistical functions
    aggregation functions, 128
    data models, 129
    DAX function, 130, 131
    SalesInfo table, 127
techniques, 103
text information
    DAX formulas, 100
    formula bar, 97
    LEFT() DAX function, 98
    modification, 96

457

© Adam Aspin 2023
A. Aspin, *Pro DAX and Data Modeling in Power BI*, https://doi.org/10.1007/978-1-4842-8995-2

R

U, V, W

X, Y, Z

Printed in the United States
by Baker & Taylor Publisher Services